T0219555

Cambridge Elements ≡

Elements of Flexible and Large-Area Electronics
edited by
Ravinder Dahiya
University of Glasgow
Luigi Occhipinti
University of Cambridge

INTEGRATION TECHNIQUES FOR MICRO/ NANOSTRUCTURE-BASED LARGE-AREA ELECTRONICS

Carlos García Núñez
University of Glasgow

Fengyuan Liu
University of Glasgow

Sheng Xu
University of California, San Diego

Ravinder Dahiya
University of Glasgow

CAMBRIDGE
UNIVERSITY PRESS

CAMBRIDGE
UNIVERSITY PRESS

University Printing House, Cambridge CB2 8BS, United Kingdom

One Liberty Plaza, 20th Floor, New York, NY 10006, USA

477 Williamstown Road, Port Melbourne, VIC 3207, Australia

314–321, 3rd Floor, Plot 3, Splendor Forum, Jasola District Centre, New Delhi – 110025, India

79 Anson Road, #06–04/06, Singapore 079906

Cambridge University Press is part of the University of Cambridge.

It furthers the University's mission by disseminating knowledge in the pursuit of education, learning, and research at the highest international levels of excellence.

www.cambridge.org
Information on this title: www.cambridge.org/9781108703529
DOI: 10.1017/9781108691574

First published 2018

A catalogue record for this publication is available from the British Library.

ISBN 978-1-108-70352-9 Paperback
ISSN 2398-4015 (online)
ISSN 2514-3840 (print)

Integration Techniques for Micro/ Nanostructure-Based Large-Area Electronics

DOI: 10.1017/9781108691574
First published online: 1 November 2018

Carlos García Núñez
University of Glasgow

Fengyuan Liu
University of Glasgow

Sheng Xu
University of California, San Diego

Ravinder Dahiya
University of Glasgow

Abstract: Advanced nanostructured materials such as organic and inorganic micro/nanostructures are excellent building blocks for electronics, optoelectronics, sensing and photovoltaics because of their high crystallinity, long aspect ratio, high surface-to-volume ratio and low dimensionality. However, their assembly over large areas and integration in functional circuits is a matter of intensive investigation. This Element provides detailed descriptions of various technologies to realize micro/nanostructure-based large-area electronic (LAE) devices on rigid or flexible/stretchable substrates.

The first section of this Element provides an introduction to the state-of-the-art integration techniques used to fabricate LAE devices based on different kinds of micro/nanostructures. The second section describes inorganic and organic micro/nanostructures, including the most common and promising synthesis procedures. In the third section we explain different techniques that have great potential for integration of micro/nanostructures over large areas. Finally, the fourth section summarizes important remarks about LAE devices based on micro/nanostructures, and future directions.

Keywords: Large-Area Electronics, Printed Electronics, Nanotechnology, Nanostructures, Microstructures, Nanowires, Nanotubes

ISBNs: 9781108703529 (PB), 9781108691574 (OC)
ISSNs: 2398-4015 (online), 2514-3840 (print)

Contents

1 Introduction

Current technology is progressing fast towards the development of low-cost and high-performance processes for the fabrication of large-area electronic (LAE) applications on both rigid and flexible/stretchable substrates, including flexible field effect transistors (FETs) [1–6], thin film transistors (TFTs) [7, 8], transparent flexible LAE devices [7, 9], flexible tactile *e*-skin [10–15], smart-coatings for photovoltaics [16, 17], flexible patches for health monitoring [18-24], flexible metal-semiconductor-metal (MSM) LAE devices [25], flexible optoelectronics [26, 27], stretchable conductive interconnects [28] and flexible integrated circuits (ICs) [29, 30]. The compatibility of LAE device fabrication processes with roll-to-roll (R2R) manufacturing and CMOS technology would allow the production cost of LAE devices to be reduced, and therefore to become affordable to a wider market [31]. Conventional LAE devices are based on either bulk or thin film materials [12, 31, 32]. However, these materials are already obsolete in terms of the requirements of current electronics (lower power consumption, higher efficiency electrical transport, faster switching speeds...), and therefore, novel advanced nanostructured materials, such as inorganic (e.g. semiconductors and metals) and organic materials with the shape of nanowires (NWs) and nanotubes (NTs), have demonstrated an excellent potential as building blocks in many LAE applications, including electronics, optoelectronics, photovoltaics, photonics and sensing [2, 4, 6, 17, 23, 33–44]. The low dimension, long aspect ratio, high crystallinity, high surface-to-volume ratio, high on-currents, high switching speeds, etc. with respect to their counterparts based on bulk or thin film materials, make filamentary nanostructures such as NWs and NTs excellent candidates to improve further the functionality of conventional LAE devices (i.e. based on thin films and bulk materials). Accordingly, organic/inorganic NWs and NTs would allow: 1) fabricating LAE devices on both rigid and flexible substrates [45]; 2) overcoming state-of-the-art planar CMOS technology by scaling on-currents and switching speeds [46]; 3) creating three-dimensional (3D) integrated circuits (ICs) [33]; and 4) continuing the Moore's law due to their higher integrability [47]. However, one of the most important challenges still facing the bottom-up fabrication approach of LAE devices based on nanostructures is the integration and assembly of these nanostructures at well-defined locations over large-area substrates, and with high reproducibility and uniformity. Although many efforts have been expended over recent years, the above drawbacks have not been fully addressed so far.

The fabrication of LAE devices based on nanostructures such as NWs and NTs requires not only high alignment accuracy along the surface of the receiver substrate (e.g. >98% NW directional alignment) [37], but also good uniformity of the resulting electrical properties (e.g. 10 NWs/μm and 708 NWs/mm^2) [48, 49]. Over the past few years, many efforts have been dedicated to developing high-performance techniques that allow ordered arrays of nanostructures to be deposited on any kind of substrate, including flexible and rigid substrates over large areas. Focusing on the large-area integration of nanostructures on both rigid and flexible substrates, we have compiled the most promising techniques reported in the literature in terms of transfer yield and resulting device performance. Accordingly, we have divided the integration techniques into three main groups, comprising printing, assembly and lithography-based integration techniques as shown in the schematic diagram of Figure 1.1. On the one hand, printing techniques can be separated into two subgroups which depend on the use of either non-contact printing methods (e.g. inkjet printing, aerosol printing, screen-printing...) [50–52] or contact-printing methods (contact-printing, roll-printing, stamp-printing...) [37, 53, 54]. Non-contact printing techniques comprise the use of pastes or inks where nanostructures are conventionally embedded and transferred to a receiver substrate. In these conditions, the nanostructures are not in direct contact with the receiver substrate, and it is the medium (i.e. paste, ink...) that is responsible for the alignment and compactness of the resulting nanostructure layer formed on top of the receiver substrate surface. On the other hand, contact-printing techniques differ between one another in the geometry of the donor and receiver substrates (planar, cylindrical, etc.), and in this case, the transfer

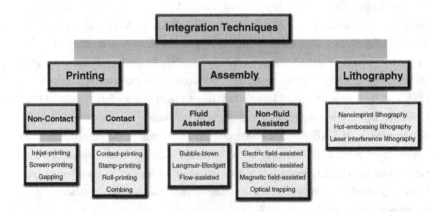

Figure 1.1 Block diagram presenting conventional and novel techniques utilized to integrate micro- and nanostructures over large areas on both rigid and non-conventional flexible substrates.

mechanism comprises the direct contact of the nanostructures with the receiver substrate. The above printing-based techniques have demonstrated their potential and scalability towards large areas – wafer-scale integration of nanostructures – allowing the successful fabrication of LAE devices based on nanostructures and improving nanostructure transfer-yields and process reproducibility. In this regard, printing techniques are a promising near-future approach for nanostructure-based devices, mainly due to their simplicity, low processing temperatures, suitability for large-area and mass production (compatibility with R2R technology), compatibility with two-dimensional (2D) and 3D monolithic integration, reproducibility, reliability and compatibility with flexible substrates. These integration techniques have already demonstrated their validity for developing multi-NW-based LAE devices on both conventional and non-conventional substrates for *e*-skins, integrated circuits (ICs), and high-efficiency interconnectors for the processing of digital information, energy harvesting and storage, consumer electronics, etc. [11, 47].

Assembly techniques mainly use nanostructures in solution, allowing nanostructures to have a 'free' motion in a liquid environment, permitting the use of electric fields (dielectrophoresis or DEP) [55], electromagnetic fields (optical and optoelectronic tweezers) [56], magnetic fields [57], fluidics [58], Langmuir–Blodgett (LB) [59] or bubble-blown (BB) [36] techniques to align – and in some cases to assemble – nanostructures at specific places on the receiver substrate. However, being wet-based, the above techniques present high material wastage and require high-cost and complex systems, hindering their use in the market for commercial applications.

Integration techniques based on lithographic methods, including nanoimprint lithography (NIL) [60, 61], hot-embossing lithography (HEL) [62] and laser interference lithography (LIL) [63] have demonstrated great potential for the development of large-area devices based on micro-/nanostructures not only on conventional rigid substrates but also on flexible non-conventional ones. In addition, the low-cost and rapid fabrication of these techniques make them promising candidates for a wide number of LAE applications based on the nanostructures discussed in this Element [64–67].

This Element shows the state-of-the-art research progress of large-area integration of nanostructures such as organic and inorganic NWs and NTs on both rigid and flexible substrates and their promising functionality in LAE devices. We will discuss the advantages and disadvantages of different integration techniques depending on the properties of the nanostructures, including their compatibility with R2R technology and scalability over large areas. The Element is organized as follows. Section 2 presents the most relevant organic and inorganic NWs and NTs that conventionally integrate in LAE devices, including common synthesis methods and their most notable properties. Section 3 summarizes printing- and non-printing-based techniques used for the large-area integration of the above nanostructures. Herein, we will emphasize the importance of preserving the properties of as-grown nanostructures during each transfer step, as well as the fabrication of high-quality electrical contacts in LAE devices. Finally, we will conclude by comparing nanostructure-based LAE devices and our future view of coming research avenues for LAE based on novel nanostructures.

2 Nanostructures

2.1 Inorganic Nanostructures

Metal and semiconductor NWs and NTs have been successfully grown from a wide selection of materials, including noble metals (Au, Ag, Cu), metal oxides (ZnO, CuO), nitrides (GaN, Si_3N_4), phosphides (GaP, InP), carbides (SiC), compound semiconductors (InAs, GaAs, CdS) [68–70] and elementary semiconductors (Si and Ge) [71, 72]. For the sake of simplicity, we have classified these nanostructures in two main groups, including metal and semiconductor materials. In this regard, this section will show conventional mechanisms and systems used to synthesize high-crystalline and uniform filamentary inorganic nanostructures with the shape of NWs and NTs, and their most relevant properties and potential applications depending on the material characteristics.

2.1.1 Semiconductor Nanowires

Semiconductor NWs are filamentary nanostructures with high aspect ratio and high surface-to-volume ratio that have been successfully synthesized by bottom-up (Figure 2.1) and top-down (Figure 2.2) approaches [73, 74], depending on their composition and structure, comprising mainly vapour–liquid–solid (VLS) [72, 75–83], vapour–solid (VS) [84], vapour–solid–solid (VSS) [85], solid-liquid-solid (SLS) [86], chemical vapor transport (CVT) [55, 87] and metal-assisted chemical etching (MACE) [17, 88]. The above mechanisms allow semiconductor NWs to be obtained that consist of single elements (Si, Ge) or compounds (III–V, metal oxides, nitrides, phosphides, etc.) with a high degree of control over properties such as crystal quality, aspect ratio, morphology and doping type/level, which are required attributes for high-performance electronics, e.g. electronics,

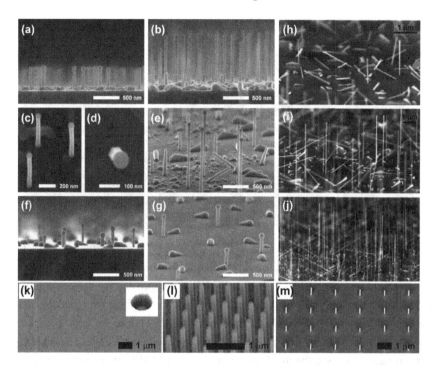

Figure 2.1 (a–g) Scanning electron microscopy (SEM) images of GaAs NWs grown by Ga-assisted VLS synthesis procedure in CBE system. Reprinted with permission from García Núñez et al.[77] Copyright © 2013, Elsevier B.V. (h–j) SEM images of GaAs NWs grown by Ga-assisted VLS on pre-patterned substrates. Reprinted with permission from Russo-Averchi et al.[115] Copyright © 2012, Royal Society of Chemistry. (k–m) SEM images of InAs NWs grown by selective area epitaxy (SAE). Reprinted with permission from Hertenberger et al. Copyright © 2010, AIP Publishing.

sensors and photovoltaics [24, 41], UV photodetectors [55, 84, 89–91], high mobility FETs [2, 6], flexible electronics [4, 6], multifunctional electronics [33] and single electron-based transistors [92].

The most extended and successful systems used to carry out the synthesis of semiconductor NWs through the aforementioned approaches are chemical vapour deposition (CVD) [93], laser ablation cluster formation [83], metal-organic chemical vapour deposition (MOCVD) [73], molecular beam epitaxy (MBE) [75, 76, 78] and chemical beam epitaxy (CBE) [77, 94]. In addition to a pure phase NW [77], the above techniques have demonstrated tremendous potential for carrying out nanoengineering along with the NW structure, morphology and composition, enabling e.g. the fabrication of

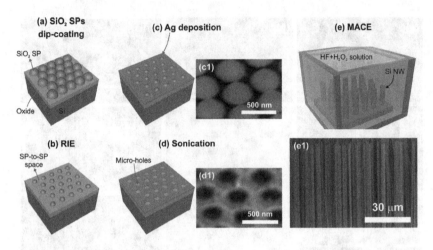

Figure 2.2 3D schema of top-down approach, namely MACE, showing experimental steps for the synthesis of Si NWs, comprising: (a) dip-coating of SiO$_2$ micro-spheres (SPs) over a large-area Si wafer surface; (b) reactive-ion etching (RIE) of SiO$_2$ SPs; (c) deposition of 100–200 nm of Ag layer (see the SEM image in [c1]); (d) sonication in isopropanol for 5 min creating a periodic metal nanomesh (see the SEM image in [d1]); (e) MACE process dipping the sample in a HF/H$_2$O$_2$ solution for 30 min, resulting in vertically aligned Si NWs (see the SEM image in [e1]). Reprinted with permission from García Núñez et al.[17] Copyright © 2018, American Chemical Society.

core–shell structures [95], superlattices [96] and polytypic and twinned structures [97].

One of the key aspects of successful LAE device manufacturing based on nanostructures is the current possibility to synthesize high crystal quality NWs over large areas. For example, in either top-down [17] or bottom-up approaches [77], the use of metal and dielectric nanoparticles over large areas for VLS (Figure 2.1) and MACE synthesis (Figure 2.2), respectively, have allowed the synthesis e.g. of III–V NWs (Figure 2.1[a–g]) [77, 98], metal oxide NWs [55, 84, 99, 100] and IV NWs (Figure 2.1[h,i] [72], Figure 2.2[e1] [17]) at wafer scale.

Group III–V Semiconductor Nanowires

The growth of III–V semiconductor compounds with the shape of NWs has been successfully demonstrated for arsenides (III-As) [98, 101], phosphides (III-P) [102, 103], nitrides (III-N) [104, 105] and, more rarely, for antimonides

(III-Sb) [106]. The possibility to grow these semiconductor materials makes this kind of NWs attractive for the development of many applications, including electronics, optoelectronics, photonics, photovoltaics, sensing, etc. [107–109]. III–V semiconductor NWs are conventionally grown in CBE, MBE and MOCVD systems by the well-known metal-assisted VLS mechanisms [110]. Au is typically used as the metallic catalyst to guide the growth of the NWs, mainly due to the possibility of forming an eutectic with most of the III–V elements involved in the NW growth [111]. For that, Au nanoparticles (NPs) are typically formed on top of the growth substrate mainly through two different strategies, involving: i) spin-coating (or drop-casting) of Au colloidal NPs covering the surface of the growth substrate [112], or ii) de-wetting of Au thin films to produce Au NPs [113]. Alternatively, III–V NWs have also been grown by processes such as self-catalysed VLS, comprising the use of the element III (i.e. Ga, In, etc.) as the catalyst of the VLS growth. For example, the Ga-assisted VLS growth of high crystal quality GaAs NWs with pure zinc blende structure has been demonstrated (Figure 2.1[a–g]) [78, 81]. Self-assisted methods prevent the use of foreign materials such as Au, which can hinder or affect the resulting electronic and optoelectronic properties of the NW due to the intentional incorporation of Au into the III–V structure [114]. The advances demonstrated in lithography techniques that allow the definition of metallic NPs with great control over features such as NP size, NP position, NP-to-NP spacing, NPs density, etc. have fostered the scaling of III–V NW growth towards wafer scales [115]. The utilization of pre-patterned growth substrates, typically consisting of Si(111) substrates covered by a thick layer of SiOx and nano-holes predefined along its surface, has made Ga-assisted growths a promising method to achieve the large-area growth of III–V NWs (Figure 2.1[h–j]) [115].

On the other hand, III–V NWs have also been grown by catalyst-free processes [116]. The above advances in high-performance lithography methods permit great control over the dimensions of the NW (diameter and length), limiting the growth of the NW at specific locations distributed along the growth substrate surface (Figure 2.1[k–m]). Typically, nanometric holes are defined by electron beam lithography on thick oxide layers, which inhibits the nucleation of III–V on the oxide while promoting the preferential nucleation of III–V only in the defined nanopores (Figure 2.1[j]). This technique – namely, selective area epitaxy (SAE) – is one of the most promising approaches for the growth of III–V NWs at wafer scale, and therefore attractive for the development of LAE. The absence of any catalyst in the growth process is considered a key advantage of the method, preventing the formation of unintentional defects along the NW structure and therefore improving the quality of the resulting NW through a pure phase crystal. The high level of control over the position and density of NWs over the growth substrate is also considered one of the most compatible features with the techniques described in Section 3 for the development of LAE. The possibility of growing free-standing NWs with a tuneable NW footprint dimension and specific location (Figure 2.1[k,l]), increases the applicability of as-grown NWs to a wider number of circuit layouts.

Group IV Semiconductor Nanowires

IV group semiconductors stand out over the other semiconductor materials, because of their compatibility with both bottom-up [110] and top-down synthesis approaches [17, 88]. Although III–V and metal oxide semiconductors have been grown by top-down methods [117], the crystal quality of the resulting NWs is still far from the results reported for IV group NWs (Figure 2.2[e1]) [17]. The great compatibility of IV group NWs with top-down synthesis approaches is based on the mechanism governing the synthesis process. In MACE, the NWs are synthesized from the wafer, which makes this process scalable towards wafer level fabrication of free-standing NWs. In this scenario, the NWs take the crystal structure, doping level and preferential orientation of the original wafer. For example, the MACE synthesis of Si NWs demonstrates the possibility of growing Si(100) NWs vertically aligned on a Si(100) substrate through the steps described in Figure 2.2(e1). One interesting approach that allows the MACE synthesis over large areas comprises the use of metallic nanomesh to mask specific areas of the substrate, exposing the rest to the etching solution. For example, SiO_2 colloidal NPs have been dip-coated on the surface of a Si(100) substrate (Figure 2.2[a]) to create a nanometallic mesh that was used as a catalyst in MACE synthesis of Si NWs (Figure 2.2[b–d]) [17]. In this scenario, the pore size (Figure 2.2[d1]) can be roughly controlled by the initial size of the SPs. The fundamental principle of MACE consists in the use of an etching solution (e.g. HF/H_2O_2 for Si wafers) promoting the preferential etching of the Si wafer underneath the Ag layer (Figure 2.2[e]) and resulting in vertically aligned Si NWs on top of the Si wafer (Figure 2.2[e1]). Therefore, this is a promising low-cost and easy-to-develop approach for improving the quality of NW samples for their integration over large areas through the techniques explained in Section 3.

Group IV semiconductor NWs have also been synthesized successfully by bottom-up approaches such as Au-assisted VLS. Indeed, as stated previously, VLS theory was firstly proposed to explain the growth of Si NWs on Si substrates using Au nano seed as catalyst (Figure 2.3[a]) [118]. After its first statement, Au-assisted VLS has been further researched, demonstrating the

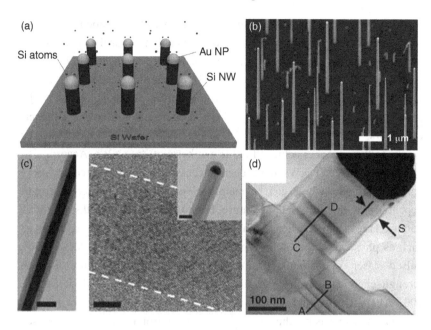

Figure 2.3 (a) 3D schema and (b) SEM image of Au-assisted VLS Si NWs. Reprinted with permission from Hannon et al. [72] Copyright © 2006, Springer Nature. (c) SEM image of i-Si/SiO$_x$/p-Si core–shell NWs. Reprinted with permission from Lauhon et al. [120] Copyright © 2002, Springer Nature. (d) Transmission electron microscopy (TEM) image of Si/Ge longitudinal heterostructures grown along a Si NW by VLS. Reprinted with permission from Zakharov et al. [122] Copyright © 2006, Elsevier B.V.

growth of longer free-standing NWs made of Si (Figure 2.3[b]) [72], Ge [71, 119], and even the fabrication of heterostructures based on these materials, including Ge/Si, i-Si/SiO$_x$/p-Si, i-Si/p-Si core/shell structures (Figure 2.3[e]) [120, 121] and longitudinal heterostructures (Figure 2.3[d]) [122] with the shape of a NW.

Metal Oxide Semiconductor Nanowires

Among all metal oxide materials that can be synthesized with the shape of a NW, ZnO is one of the most attractive metal oxide nanomaterials for the development of LAE. The growth and characterization of ZnO NWs have been thoroughly studied during decades, and therefore, extensively reported in the literature [123–126]. Due to their facile fabrication procedure and excellent properties (high surface-to-volume ratio, high surface reactivity, wide band gap, sensitivity to UV light, piezoelectric effects, biocompatibility, etc.), ZnO NWs have been considered a great candidate for the development of a wide range of applications, including sensors (gas sensors [127], photodetectors [55], and chemical sensors [128]), photovoltaic cells [129], photonics [130], energy harvesters (piezoelectric nanogenerators [131], and triboelectric nanogenerators [132]) and energy storage devices (supercapacitors, and batteries) [133]. These investigations have been fostered thanks to the advances achieved in the synthesis of ZnO NWs. Accordingly, the growth of ZnO NWs have been successfully demonstrated by using various methods, which can be mainly classified in two different groups depending on the growth temperature. High temperature growths (i.e. 500–1500 °C) are usually carried out in gaseous environments in closed chambers such as physical vapour deposition (PVD) [134], CVD [135], MBE [136], pulsed laser deposition (PLD) [137] and high-temperature furnace through chemical vapour transport (CVT) (Figure 2.4[a–c]) [55, 84, 99, 100]. Alternatively, ZnO NWs have been grown by low temperature methods (< 200°C), comprising the use of solution environments. Solution-based growths mainly include hydrothermal methods [138–140] and electrochemical deposition [141].

Under high temperature growth conditions, ZnO NWs have been successfully grown mainly by VLS and VS mechanisms [55, 84, 87]. Although it has been rarely reported, ZnO-seeded VS growth of ZnO NWs have been demonstrated [87]. In VLS conditions, there are two accepted models – namely Gibbs–Thomson and the diffusion model – that explain the relation between length and diameter of the resulting NWs. One can clearly identify that ZnO NWs are growing through the Gibbs–Thomson model if the characterization of the resulting NWs presents an increase in the NW length when the diameter of the NW increases [142]. In this scenario, the adatoms are directly attached to the NW tip (solid phase) after diffusing through the liquid metal catalyst located at the tip of the NW. In contrast, when the growth of the NW is governed by the diffusion model, the NW growth is limited by the diffusion of adatoms along the NW sidewalls, resulting in NW lengths inversely proportional to the NW diameter [143].

In metal-assisted VLS of ZnO NWs, governed either by Gibbs–Thomson or diffusion model, Au nanoparticles are typically used as catalyst (Figure 2.4[c]), acting as a trap for growth species, i.e. Zn and O_2 gases. The supersaturation of the Au catalyst produces a precipitation of ZnO crystals to the interface formed between the catalyst and the substrate surface. ZnO crystals nucleate under the catalyst and on the substrate

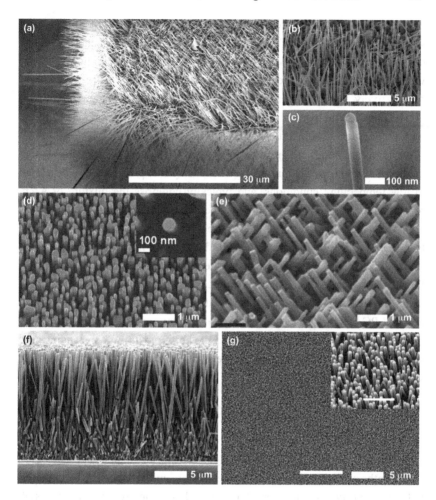

Figure 2.4 (a–c) SEM images of ZnO NWs grown by CVT on sapphire substrates. (c) Au is commonly used as a metal catalyst (see the Au nanoparticle at tip of the NW) to guide the growth of ZnO NWs along a preferential direction fixed by the crystal orientation of the substrate surface (i.e. sapphire) through the well-known VLS growth mechanism. Reprinted with permission from García Núñez et al.[99] Copyright © 2018, Springer. (d) Vertically aligned ZnO NWs grown by CVT through VS method on *c*-plane sapphire substrate. Reprinted with permission from Shen et al.[87] Copyright © 2013, AIP Publishing. (e) ZnO NWs grown 30° tilted with respect to *m*-plane sapphire substrate surface. Reprinted with permission from Hou et al. [145] Copyright © 2003, AIP Publishing. (f) Cross-section and (g) top-view SEM images of ZnO NWs grown by liquid-phase chemical process. Reprinted with permission from Xu et al. [129] Copyright © 2010, American Chemical Society.

surface, following the preferential orientation fixed by the crystalline structure of substrate surface. In this regard, substrates such as c-plane sapphire and Si(111) are suitable for the growth of free-standing ZnO NWs vertically aligned on the substrate surface (Figure 2.4[a–d]) [144]. Alternatively, other conventional substrates such as m-plane sapphire or Si(100) have demonstrated their suitability for producing ZnO NWs titled with respect to the substrate surface (Figure 2.4[e]) [145].

On the other hand, the growth of ZnO NWs by hydrothermal solid-based methods, also known as liquid-phase chemical (LPC) processes, is carried out in low temperature conditions, and typically comprise the mixture of a Zn precursor such as zinc nitrate and hexamethylenetetramine (HMTA) in water [134, 146–149]. From the analysis of the ZnO NW synthesis kinetics, it has been found that high crystal quality ZnO NWs result from slow hydrolysis of HMTA, producing hydroxyl groups to the solution. The key to the ZnO NW formation is due to the dehydration process resulting from heating the zinc hydroxide reaction [147–149]. Compared to gas phase-based synthesis processes, LPC highlights because of its scalability towards wafer-scale growth of ZnO NWs [129]. LPC has demonstrated the ability to grow free-standing and high crystal quality ZnO NWs with great aspect ratios (>100) and high densities (Figure 2.4[f]) over large areas (Figure 2.4[g]).

2.1.2 Semiconductor Nanotubes

Semiconductor NTs derived from various materials, including C, BN, MoS_2, SiO_2, GaN, WS_2, AlN, In_2O_3, V_2O_5, $H_3Ti_2O_7$, etc. [150], have been successfully synthesized through the technique, namely epitaxial casting, and thanks to the possibility of creating epitaxial core–shell structures. The synthesis of core–shell NWs, and subsequent dissolution of the inner core, have permitted the creation of single-crystalline NTs based on the above-named semiconductor materials. The validity of this approach has been demonstrated in those cases where the chemical stability of core/shell materials is different. For example, the fabrication of GaN NTs with inner diameters of 30–200 nm and NT wall thickness of 5–50 nm has been reported [151]. ZnO NWs have been successfully used as core (i.e. hard template) for creating GaN [150]. Right after the MOCVD deposition of GaN coating ZnO NW surface, the core ZnO NW was removed by simple thermal reduction and evaporation in an NH_3/H_2 environment, resulting in an ordered array of GaN NTs on the growth substrate [150]. Silica NTs have also been synthesized using Si NWs as a core template, showing high control over the resulting NT wall thickness. The opportunity to oxidize Si NWs through thermal oxidization processes, and a subsequent wet etching of the Si NW-based core, allows the creation of silica NTs as well [152]. Further reports have shown the possibility of creating metal nitrides and metal oxide-based NTs, e.g. AlN [153] and In_2O_3 [154], respectively. However, the formation mechanisms governing the creation of the NT structure in the above examples are not as well controlled as in the case of epitaxial casting [150]. For that reason, further investigations will be needed to reach an accurate control over the resulting NT morphology and structure.

One of the most promising high aspect ratio nanostructures for LAE applications on both rigid and flexible substrates is carbon nanotubes (CNTs), being tubular carbon-based nanostructures which can be understood as single-wall (SWNTs) or multi-wall (MWNTs) of carbon atoms rolled up into a seamless cylinder. CNTs present unique properties such as high intrinsic carrier mobility and conductivity [155], mechanical flexibility [156] and low-cost mass production [157, 158]. Conventional applications are based on thin film CNTs rather than single CNTs to reduce the variability of CNTs' electrical properties produced by their chirality and diameter. CNTs have been grown mainly by arc-discharge, laser ablation and CVD techniques [159]. The existence of the above high-performance systems to grow or regrow selectively SWNTs with a predetermined chirality [160–163] fosters their applicability in LAE.

Unlike their carbon-based allotrope, gapless two-dimensional graphene, SWCNTs can be either semiconducting or metallic, depending on their chiral vector (n, m). This variety in electrical properties remains one of the main hurdles confronting their large-area application, especially when it comes to electronics, optoelectronics and circuit realization. Consequently, highly selective sorting is needed. At present, different means of solution have been demonstrated and are mainly categorized into two groups, including selective growth or post-synthetic sorting [164]. The former usually refers to the CNTs synthesized by the CVD method with well-controlled growth conditions such as catalyst size, precursors flow, substrate temperature and plasma environment (residual pressure, and gas content/concentration), to obtain high structural and electrical properties [165–167]. On the other hand, the post-synthetic sorting-based growth method takes advantage of different chemical (through selective function) [168, 169], physical (through density-gradient ultracentrifugation) [170, 171] or electrical (through current-induced electrical breakdown) [172] properties of semiconducting and metallic SWCNTs to differentiate them.

Thanks to those pioneering works in fields relating to large-area synthesis and sorting of CNTs, various LAE applications using CNTs as building blocks have reached levels approaching industrialization. TFTs [155], sensors and photodetectors [23, 173], touch screens and displays [174, 175], and conducting interconnects [176], are some examples of successful utilization of CNTs as active/passive components. Preliminary works showed that a random network based on semiconducting and metallic CNTs is a potential approach to develop LAE [177]. The ease of development and cost-effective method of fabricating a network of CNTs, as well as the compatibility of this configuration with the doping process and non-conventional substrates, could be an opportunity for these random networks of nanostructures to impact LAE. In this Element, we will also highlight the best performance integration techniques used to transfer and to align CNTs from growth substrate to receiver foreign substrate for the fabrication of high-performance LAE applications.

2.1.3 Metal Nanowires

Various methods have probed the viability of synthesizing gold (Au) NWs, including solution phase reduction of $HAuCl_4$ in a micellar structure formed either by cetyltrimethylammonium bromide (CTAB) in an aqueous solution [178] or by oleic acid/oleylamine in an organic solvent [179], and reduction of $HAuCl_4$ in a nanoporous anodic aluminium oxide (AAO) template [180] (or other kind of template) [181]. Moreover, Au NWs have been fabricated by physical deposition of Au on a lithographically patterned substrate [182]. Characterization of the resulting Au NWs shows that these nanostructures are chemically inert, have very low resistivity and present a strong plasmonic effect, which makes them excellent candidates as interconnects for linking molecular devices [182], building blocks for sensing applications [183] and nanometric gratings for optical polarizers [184].

The ability to grow uniform and long-aspect-ratio silver (Ag) NWs, demonstrated in recent decades, has promoted the extensive utilization of Ag NWs in a wide range of LAE applications, including transparent/flexible electrodes [185], biosensors [186], transparent thin film heaters [187], light-emitting diodes (LEDs) [188] and touch screens [40]. In the early stages of investigating the synthesis of Ag NWs, researchers used electrochemical methods based on hard templates (nanoporous membranes [189] and CNTs [190]) or soft templates, resulting in non-uniform NWs and low-synthesis yields. From those studies, hard-template methods showed more reproducibility and controllability over the resulting morphology of the Ag NWs in contrast to soft-template approaches. On the other hand, soft templates such as surfactants, micelles and different polymers have been found to overcome the drawbacks of hard templates, mainly due to the key ability of soft templates to be dissolved in solution [191–193].

2.1.4 Metal Nanotubes

Metal NTs are mainly obtained by depositing a uniform metallic layer either atop the surface of a sacrificial NW (or NT) that is removed after the deposition, or filling nanoporous membranes, resulting in a nanostructure with the shape of a tube. For example, Pd NTs using ZnO as sacrificial NWs [194], and Co, Pt, and CoPt NTs using AAO membranes [195]. Alternatively, metal NTs have been obtained by creating a uniform metallic layer atop the surface of CNTs, benefited by the strong interaction of some metals and carbon. However, not all the metals exhibit such strong interaction with the CNT surface [196]. A valid strategy to extend the suitability of metals for the above method relies on the use of Ti, Cr or Ni layers to form an interface between the CNT surface and the metal coating [196].

One of the main advantages of metal NTs compared to NWs (that also applied to semiconductor NTs) is the significant increase in the sensitive surface of NTs with respect to NWs, which makes them excellent candidates for the development of LAE applications [197].

2.1.5 Quantum Dots

Quantum dots (QDs) are zero-dimensional (0D) structures which have been proved with several promising applications in modern electronics. Unlike the traditional CMOS process, which requires the use of expensive equipment and a delicate process flow, processing quantum dots (QDs) in the format of thin film for electronics application is simple, cost-effective and holds great promise in manufacturing. Various dry- and wet-based methods have been attempted to obtain the desired thin film, including drop-casting [198], spin-coating [199–202], transfer-printing [203-206] and inkjet-printing [207]. Among them, drop-casting and spin-coating are the two widely used techniques for the thin film process for a quantum dot-based solution. With careful control of the volume of solution used in the process and the area of sample to be coated/casted, a uniform and large area ($>50 \times 50$ cm^2) quantum dot-based thin film can be achievable (Figure 2.5[a]) [198]. In order to obtain QD-based film in the desired pattern, a transfer-printing method with adhesive stamps (usually poly(dimethylsiloxane) (PDMS) stamp) was developed. The stamp can be either structured or non-structured (planar). In the former scenario, the pattern is readily defined by the structure extrusion in the stamp used (Figure 2.5[b])[203]. However, the resolution of the pattern is limited, with some discrepancy between the original design

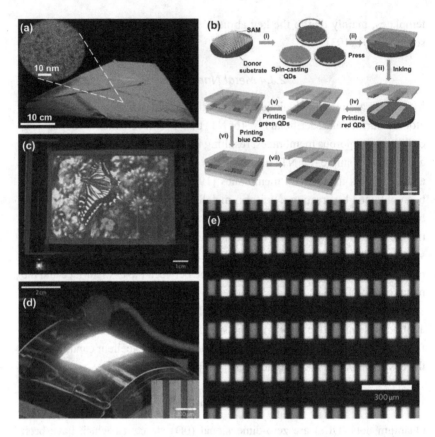

Figure 2.5 (a) Optical and SEM images (inset) of flexible and freestanding films of InP/ZnS QDs integrated over large areas (>50 × 50 cm^2). Reprinted with permission from Mutlugun et al. [198] Copyright © 2012, American Chemical Society. (b) 3D schematic of transfer-printing process for patterning of QDs. (c) Electroluminescence image of a 4-inch full-colour QDs based display using a HIZO TFT backplane with a 320 × 240 pixel array. (d) Flexible LED with RGB-based QDs patterned by transfer-printing technique. (e) Optical image of simultaneous RGB electroluminescence emission from an active matrix drive with transfer printed RGB QDs. Reprinted with permission from Kim et al. [203] Copyright © 2011, Springer Nature.

and the resulting sample [204, 205]. To solve this problem, a so-called intaglio transfer-printing method with planar PDMS stamp was developed [206]. Compared to the former transfer-printing method, this strategy enables a much more precise shape to be maintained during printing, which is crucial for high-resolution pattern transfer, allowing the fabrication of LAE such as displays (Figure 2.5[c]), flexible LEDs (Figure 2.5[d]) and an active matrix of RGB based on QDs (Figure 2.5[e]). Besides the transfer-printing, inkjet-printing is another method for QDs deposition [207]. This approach holds the benefits such as low material consumption and ease of operation, which is naturally suitable for large area electronics (LAE) technology.

So far, applications on several aspects have been demonstrated with QDs over large areas. For example, QDs can function as channel material for FET applications. The type and density of the transported carriers is controlled by the core of the QDs, while the doping and energy band modification can be achieved by the modification of the surface of QDs [208]. By tuning the gate voltage, the major carriers in the film formed by the QDs can be controlled, and can thus switch the transistors between on and off state. Moreover, flexible large-area devices and circuits have been successfully demonstrated with spin-coated film of QDs, showing performance on a par with organic and carbon nanotube-based circuits [209, 210]. Another major field in which QDs play an important role is optoelectronic devices such as photodetectors, photovoltaic cells and electroluminescent devices. The employment of the QDs can lead to maximal radiative efficiency, which enhances the figure of merit of the optoelectronic devices [211, 212].

2.1.6 Two-Dimensional Materials

Since the discovery of graphene in 2004 [213], the 2D materials covering semi-metals [213], semiconductors [214-216] and insulators [217] with different electronic and optoelectronic properties have been explored for various applications. So far, prototype devices made from 2D materials have shown great promise by outperforming the mature existing technologies [218–220]. Therefore, how the 2D material will fit into the electronic industry based on CMOS technology is one area of interest in the post-silicon era.

Meanwhile, LAE – including flexible and printable electronics – is a new way of making and using electronics over large areas with optimized input. In this regard, 2D materials are considered to hold great promise due to their unique properties such as high flexibility [221, 222], high transparency [223] and extraordinary electronic property [224], as well as the benefit of cost-effective material synthesis over large quantities [225]. Known for more than a decade, different strategies including mechanical exfoliation [213], molecular assembly [226], liquid phase exfoliation [227–229], epitaxial growth [230, 231] and CVD method [225] have been realized for 2D material synthesis. The first approach to be discovered is the so-called mechanical exfoliation method from bulk crystal. While this method offers the highest material quality and superb surface cleanness, it is so far limited to laboratory scale and is not promising for large-area production in the near future [213]. By contrast, methods of liquid phase exfoliation can provide the highest throughput; the quality of the material is thus limited and high-performance device applications are not favourable. If we consider the balance between input (cost) and output (quality of the product), the CVD method is regarded as one of the most promising approaches for large-area application. Demonstrated in 2009 (the synthesis of monolayer graphene [225]), this method has now been extended to the synthesis of diverse kinds of 2D materials [232–234].Taking graphene as an example, the CVD process involves using copper as the catalyst with a flow of methane as the carbon source. In the reduction atmosphere (a mixture of Ar and H_2) and high temperature, monolayer graphene can be grown on the surface of the copper foil [225]. The size of the graphene synthesized by this method is not a problem; the only limitation comes from the size of the metal foil used as the catalyst. During the preliminary stage, the synthesized graphene is polycrystalline with an average domain size around several microns [225]. After several years' efforts, single crystal graphene with large domain size was successfully realized by following this approach [235, 236], with the domain size up to centimetre [237]. Although the synthesized condition is rigorous, with a high temperature and accurate control of inlet gases, this process still holds great promise for industry-level production towards single-crystal electronics. Surprisingly, with several years' effort, the 2D materials synthesized by CVD method have shown electronic properties comparable to the mechanical exfoliated samples, with device mobility up to 30,000 $cm^2/V\cdot s$ at room temperature [237]. Controlling the cost is another concern in 2D material synthesis. In this regard, efforts have been made to significantly decrease the cost of a single CVD process [27]. After synthesis, there is a need to transfer the material to the desired substrate for further device processing and applications. The initial method involves employing a thin Poly(methyl methacrylate) (PMMA) layer as the mechanical support for the transfer process [225]. However, drawbacks of this method, such as wrinkles in the transferred film and recalcitrant surface contaminations, have perplexed the researchers for several years. After several pioneering attempts [238–240], people realized that this type of polymer tends to leave a strong residue on a graphene surface due to its reaction with the etching solution [241]. Various replacement polymers have been proposed and demonstrated to lead to a much cleaner surface [241]. Despite the bulky and critical process flow of material synthesis and transfer before realizing a real application, R2R synthesis and transfer technique have been successfully realized for flexible touch screens. This is a vivid example revealing the potential of the 2D material in LAEs [242]. Meanwhile, another interesting strategy to transfer graphene to the desired substrate is by using a hot pressing method [27]. Assisted by heat and pressure, the graphene can be transferred successfully to both flexible and rigid substrates after releasing the film from the metal foil by using thermal release tape (Figure 2.6[a]) [27, 243]. This method has been further expanded to the development of e-skin based on single-layer graphene for robotics [11].

2D materials synthesized on non-metal substrate (such as silicon) is another approach which has been explored for the large-area application of graphene [244, 245]. Recently, researchers have realized wafer-scale single crystal graphene synthesis on hydrogen-terminated germanium substrate by taking advantage of unidirectional alignment of multiple seeds and stitching them together [246]. More interestingly, the weak interaction between the as-synthesized material and the underlying substrate allows for easy transfer without any etching, greatly simplifying the process flow for graphene electronics, and thus the cost is reduced. Another interesting method that can achieve synthesis and transfer up to wafer scale on a silicon wafer is the face-to-face method (Figure 2.6[b]) [247]. Benefiting from the capillary force generated by the plasma-induced bubbles, this method provides simple transfer of the film without any cracking or wrinkles, as demonstrated by atomic force microscopy (AFM) (Figure 2.6[c]).

Besides the touch screen application, which has been pointed out, diverse large-area applications have been realized with CVD-synthesized 2D materials. For example, large-area FET and circuit fabrication have been demonstrated with an acceptable value of figure of merit of devices such as mobility and contact resistance [248, 249]. Besides, large-area, transparent and stretchable electrodes [250] (Figure 2.6[d,e]), RF devices [251], photodetectors [252, 253] and pressure and chemical sensors [254–256] have been successfully realized with 2D materials, with various unique and interesting features.

Another interesting method that must be pointed out for large-area 2D material synthesis is laser scribing [257]. With the direct laser reduction effect on the graphene oxide (GO), the film can be reduced for electrochemical capacitor and antenna applications [258]. While this method suffers from low quality in the reduced film, another so-called microwave reduction method [259] is able to successfully convert GO film to high-quality graphene. The resulting material exhibits high crystallinity with a comparable

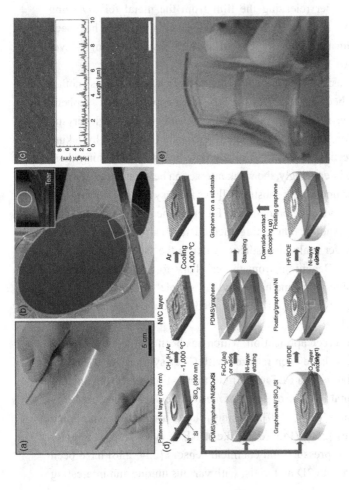

Figure 2.6 (a) Photograph of the graphene transferred on Polyvinyl chloride (PVC) substrates by using hot lamination method. Reprinted with permission from Polat et al. [27] Copyright © 2015, Springer Nature. (b) Photograph of face-to-face transferred graphene on 8-inch and 4-inch wafers. (c) AFM analysis of large-area transfer of graphene. Reprinted with permission from Gao et al. [247] Copyright © 2013, Springer Nature. (d) 3D schematic of large-area synthesis of graphene on Ni and transfer by PDMS-based method. (e) Photograph of graphene transferred to a transparent, flexible and stretchable substrate. Reprinted with permission from Kim et al. [250] Copyright © 2009, Springer Nature.

quality to the CVD synthesized film, as confirmed by Raman spectroscopy and TEM characterization. And the application of such high-quality graphene obtained by microwave reduction will be the same as previously pointed out for the graphene grown by the CVD method.

2.2 Organic Nanostructures

Organic NWs have mainly been obtained from self-assembly of organic semiconductor molecules and conducting polymers, and are considered one of the main counterparts of inorganic NWs for near-future LAE devices and circuits. Compared to inorganic NWs, organic NWs provide advantages such as rapid processing, low-cost synthesis procedures [260] and a large abundance of organic materials in earth (namely earth-abundant materials), allowing wider electronic properties to be covered in developing different LAE applications [261]. Since device performance based on organic NWs is strongly influenced by the NW assembling properties, the synthesis of organic NWs is one major matter of investigation in organic-based LAE field. Some studies indicate that organic NWs exhibit higher performance than their thin film counterparts, probably due to higher levels of crystal quality [262]. Organic NWs have mainly been synthesized by bottom-up self-assembly processes, such as solution deposition [260], physical vapor transport [263] and electrospinning [264]. These synthesis processes lack accurate control over the morphology of resulting organic NWs, which still hinders the use of these NWs for LAE applications.

Electronic devices based on organic NWs have shown mobilities up to 9.7 $cm^2/V \cdot s$ [265], which are comparable to those values obtained in TFTs based on amorphous Si (a-Si), validating the electronic properties of organic NWs for many applications (smartphones, laptops, screens, solar cells, etc.). Beyond that, organic NWs also offer a natural advantage in flexible/stretchable electronics for wearable systems, i.e. higher flexibility under bending conditions compared to inorganic NWs, mainly due to their lower Young modulus (organic NWs: ~2 GPa, inorganic NWs: 20–200 GPa) [266]. In this regard, ultra-flexible transistors/circuits [266, 267], sensors [3, 268], photodetectors [269] and memory [270] based on organic NWs have demonstrated stable characteristics under dynamic bending conditions, and reliability over time, taking advantage of the intrinsic softness of organic materials. Moreover, some theoretical and experimental studies suggest that field-effect mobility of transistors based on organic NWs could be further improved under bending conditions due to a significant increase of the intermolecular interaction generated in the material structure [266].

Organic NWs are still the subject of significant investigations in order to overcome drawbacks such as brittle behaviour and their vulnerable nature in air ambient environments (i.e. rapid degradation or meta-stability) and lack of robustness, which can limit the use of some large-area integration techniques such as contact-printing, roll-printing, stamp-printing, etc. However, inkjet-printing has shown promising results for the rapid, convenient and low-cost fabrication of LAE based on organic NWs, overcoming the above drawbacks [265].

3 Large-Area Printing and Integration Techniques

The integration of nanostructures, such as NWs and NTs, into functional circuits requires a high level of control over their directional alignment, and high and uniform linear/surface density at spatially defined locations over the device surface. In this regard, the fabrication of NW-based devices requires not only a uniform NW morphology with a high crystal quality structure, well-oriented NWs, uniform NW aspect ratio and density, but also advanced integration techniques to carry out the high-performance assembly and alignment of those NWs on foreign receiver substrates (i.e. different from the growth substrate) [42]. In this regard, there is intensive research focused on high-performance techniques for the assembly of nanostructures with reproducible results in terms of alignment, density, spacing and uniformity, making this research field very interesting for the development of near-future nanotechnology over large areas.

Integration of nanostructures such as NWs and NTs over large areas has been carried out successfully by the use of various techniques, as briefly presented in Figure 1.1. We have classified these techniques into three main groups, comprising printing, assembly and lithography-based integration techniques. In this section, we present the potential of these techniques for the integration of NWs over large-area substrates.

3.1 Integration by Printing Techniques
3.1.1 Non-Contact Printing Techniques

Inkjet-printing is a well-known wet-printing technique that has successfully demonstrated potential for the integration of both NWs [50, 271] and CNTs [51, 52] over large-area substrates. This technique requires the preparation of an ink with the NWs, also called the *NWs composite*, which commonly needs chemical modification of the nanostructure surface in order to make them dispersible into a solvent. The NWs ink is directly printed on a receiver

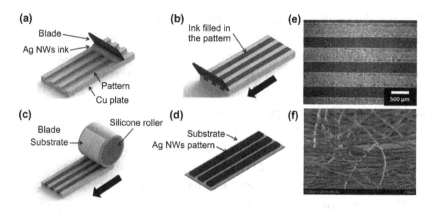

Figure 3.1 (a–d) 3D schematic illustration of conventional inkjet-printing of NWs on pre-patterned substrates. Morphological characterization, comprising (e) optical microscopy and (f) SEM, of micrometric patterns consisting of multi-NWs printed through the inkjet-printing approach. Reprinted with permission from Park et al.[271] Copyright © 2010, Elsevier Publishing Group.

substrate (Figure 3.1[a]), which can be pre-patterned in order to control the positioning of the printed NWs. A blade forming a specific angle with respect to the horizontal plane spreads the ink along the pre-pattern substrate, filling the engraved patterns with the NWs ink (Figure 3.1[b]). Finally, printed NWs can be transferred to the foreign substrate by techniques such as roll-printing (Figure 3.1[c]) or stamp-printing, which will be explained below, resulting in a multi-NW-based array as schematically described in Figure 3.1(d), and observed by optical microscopy (Figure 3.1[e]) and SEM (Figure 3.1[f]). Inkjet-printing is easy in terms of handling and storage, while fabrication involves the use of inexpensive printing cartridges and non-contact deposition onto substrates. In addition, inkjet-printing is easily scalable towards wafer-scale fabrication and has demonstrated its compatibility with R2R technology. The main drawback of inkjet-printing of NWs that has not been addressed so far is its poor control over the directional alignment of NWs and non-uniform NW density (Figure 3.1[f]), hindering the current reproducibility and applicability of this technology for the development of LAE applications based on nanostructures such as NWs and NTs.

Screen-printing is considered one of the most promising techniques for the fabrication of LAE applications, including sensors, displays, electronics, wearable systems, e-skins, smart-coatings, photovoltaics, etc. [35]

Although the validity of the application of this technique to print thick and thin films of different materials has been tested, its suitability for printing nanostructures over large areas has still not been well demonstrated. Prior to the screen-printing process, a high mass of NWs is conventionally mixed with a solvent, forming a NWs paste, typically, with a higher viscosity than the NWs ink used in inkjet-printing. Then, the NWs paste is directly screen-printed on the receiver substrate. This approach has been successfully used to fabricate devices such as humidity sensors based on ZnO NWs [272], gas sensors based on SnO_2 NWs [273], electrochemical sensors based on CNTs [274] and stretchable interconnects based on Ag NWs for wearable systems [275]. As observed in the inkjet-printing approach, screen-printing is suitable for large-area printing of different pastes, including pastes with embedded nanomaterials. However, this technique presents a significant bottleneck related to the resulting thickness of the printed layer and the lack of control over the printed NWs density and directional alignment, which reduces its current applicability for fabrication of LAE devices based on NWs and NTs.

Gapping Method has been extensively used for integrating organic NWs and nanofibers over large areas. In principle, the above nanostructures can be well aligned between electrodes by introducing an air gap into a conventional collector and applying a high voltage between the electrodes, as schematically described in Figure 3.2(a). This technique has demonstrated tremendous potential for large-area assembly of nanostructures such as NWs, NTs and organic nanofibers (Figure 3.2[b–g]). For example, the gapping method was used to assemble organic nanofibers, with a diameter below 150 nm, bridging gaps from hundreds of µm to few cm [276]. In this scenario, different materials and shapes have been successfully integrated by the gapping method, including polymer-based nanofibers [276], composite materials [277], inorganic NWs (TiO_2, SnO_2, ZnO, Ag-ZnO, $BaTiO_3$) [276] and magnetic materials [278]. Moreover, the gapping method exhibits potential for layer-by-layer fashion by rotating the collector to generate monolithic and crossed architectures (Figure 3.2[b,c]). The gapping method has been utilized to assemble organic nanofibers resulting in excellent electric and optoelectronic properties for optical polarizers [276], UV photodetectors [279] and energy storing applications [280].

In addition to solid nanofibers, NTs like semiconductor nanofibers, i.e. hollow nanofibers, have also been integrated over large areas through the gapping method. For example, TiO_2 hollow nanofibers [281], magnetic nanofibers [278], ZnO NTs and core–shell NTs [282] were assembled over large

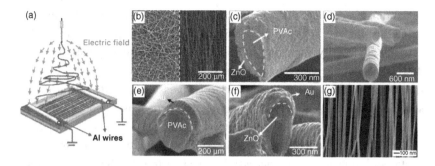

Figure 3.2 (a) 3D schematic illustration of gapping method used to integrate organic nanostructures over large areas. Fabrication of (b) uniaxial and multi-axial aligned ZnO NTs, (c) core–shell PVAc-ZnO NWs, (d) ZnO NTs, (e) core–shell PVAc-Au NWs, and (f) core–shell ZnO-Au NWs. Reprinted with permission from Choi et al. [282] Copyright © 2009, American Chemical Society. (g) Electrospinning of magnetic nanofibers of nickel ferrite over large areas. Reprinted with permission from Li et al. [278] Copyright © 2003, AIP Publishing Group.

areas with a well-controlled directional orientation allowing the fabrication of gas sensors, microfluidic devices and optical waveguides.

Therefore, one can realize the potential of the gapping method to fabricate LAE devices based on organic/inorganic NWs, NTs, nanofibers and hollow nanofibers. However, this method still presents some drawbacks that require further investigations. For example, it is observed that assembly of the afore-mentioned nanostructures over larger areas would require higher voltages to ensure the strength of the electrostatic force guiding the assembly and align-ment of the nanostructures between conductive electrodes. Accordingly, the lack of the electrostatic force produces poor alignment of the nanofibers, restricting their suitability in LAE applications.

3.1.2 Contact-Printing Techniques

Contact-printing. The principle of contact-printing involves the directional sliding of a donor substrate, consisting of a 'lawn' of free-standing NWs (Figure 3.3[a]) [54] or NTs [48, 283] on top of a receiver substrate. During the sliding step, these nanostructures tend to be aligned and combed due to the sliding shear force (Figure 3.3x[b]). Then, nanostructures are detached from the donor substrate due to the accumulation of structural strain, and finally are anchored by the Van der Waals interactions with the receiver substrate surface, resulting in nanostructures, e.g. NWs, aligned along the surface of the receiver substrate (Figure 3.3[b]). This process is initially carried out via a dry strategy, i.e. the donor substrate is gently placed upside down on top of the receiver substrates such that the NWs are in contact with the receiver substrate [33]. In this scenario, a pressure is exerted on the donor substrate from the top followed by sliding the donor substrate over a specific length range. Finally, the donor substrate is removed, resulting in a layer of NWs aligned on top of the substrate surface (Figure 3.3[c]). Dry-based contact-printing has been successfully used to transfer NWs of different materials, including Ge and Si [284], Ge/Si [33], CdSe [43] and SnO_2 [285]. Further, this approach has demonstrated a high level of control over the number of transferred NWs, allowing even the fabrication of single NW-based devices (Figure 3.3[d]), and great scalability towards large areas, i.e. exhibiting wafer-scale transfer of semiconductor NWs (Figure 3.3[e]).

The performance of the technique is further improved by using patterned [33] and/or functionalized received substrates [54], achieving better uniformity of printed NWs (NW-to-NW spacing, number of transferred NWs and length of the transferred NWs) and improvement of the selectivity of the NW placement at specific sites along the receiver substrate surface. Briefly, if the receiver substrate is previously patterned with resisting features such as those represented in Figure 3.3(f), NWs are aligned on top of both patterned lines and substrate surface. After a lift-off process, only NWs on top of the substrate surface remain, whereas the others are removed, forming well-ordered arrays of NWs (Figure 3.3[g]). This approach has been successfully used to fabricate 2D (Figure 3.3[h,i,k]) and 3D FETs, as well as arrays of photodetectors (Figure 3.3[j]), where the high accuracy obtained in placing selectively NWs makes contact-printing very attractive for the development of LAE applications.

Alternatively, contact-printing has been carried out using a spacing layer based on lubricant, which typically consists of mineral oil mixed with octane, during the transfer process, reducing the NW–NW friction [43, 54]. In this regard, the chemical binding interactions between NWs and the receiver substrate allow not only accurate control over the density of aligned NWs but also to preserve the initial length of the NWs, obtaining high NWs densities up to 8 NW/μm [54].

Contact-printing also offers an interesting feature, namely the ability to assemble and fabricate multifunctional 3D NW-based electronics on both planar and flexible substrates through monolithic printing steps [33]. In that respect, the record number of functional device layers that have been vertically stacked currently stands at 10 layers of Ge/Si NWs, stacked to form a 3D electronic structure as shown in Figure 3.3(k) [33]. The best attribute of contact

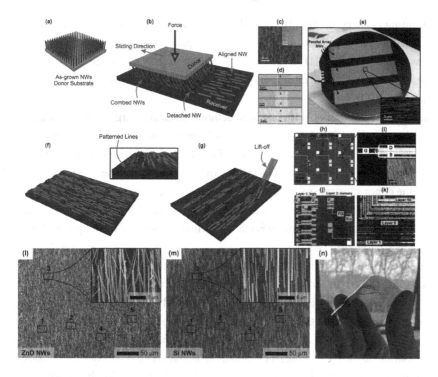

Figure 3.3 (a) Donor substrate consisting of a 'lawn' of as-grown free-standing NWs. (b) 3D schematic illustration of the contact-printing technique, showing the direct contact between donor and receiver substrates due to the applied force, and the subsequently sliding of donor along the receiver substrate surface, resulting in: i) NW combing, ii) NW detachment and, finally, NW alignment. (c) Optical image of Ge NWs (30 nm in diameter) contact-printed on PMMA lines pre-patterned on a Si/SiO$_2$ substrate (inset). (d) Parallel arrays of NWs, controlling the NW density by contact-printing conditions (i.e. from single to multi-NWs based devices). (e) Large-area and highly uniform parallel arrays of aligned Ge NWs (i.e. wafer-scale contact-printing) [54]. (f) Contact-printing on pre-patterned substrate; inset: cross-sectional zoom-in of the resultant printing. (g) Lift-off of the photolithographed lines. Reprinted with permission from Javey et al. [54] Copyright © 2008, American Chemical Society. (h,i) Optical microscope images of resultant Ge/Si NW-based films in a FET. Reprinted with permission from Javey et al. [33] Copyright © 2007, American Chemical Society. (j) Example of two-layer-based electronic circuitry based on contact-printed NWs. (k) Optical microscope image of 10 layers of Ge/Si NW-based FET. Reprinted with permission from Javey et al. [33] Copyright © 2007, American Chemical Society. SEM images of (l) ZnO and (m) Si NWs contact-printed over large areas. (n) Photograph of a flexible UV photodetector based on ZnO and Si NWs fabricated by contact-printing. Reprinted with permission from García Núñez et al. [99] Copyright © 2018, Springer Nature.

printing is the compatibility with monolithic 3D integration, which means that layer-by-layer assembly does not alter the properties of existing layers. In essence, using this 3D printing methodology with NWs based on different materials could lead to ultra-high-performance 3D electronics not accessible by scaled CMOS.

Contact-printing of NWs has been demonstrated over large areas, i.e. wafer-scale assembly of NWs [54]. In this regard, the growth of NWs over large areas is crucial to the successful realization of contact-printing also over large areas. Since the contact-printing mechanism explained above is based on the sliding of the donor substrate over the receiver, a large-scale donor substrate consisting of a lawn of NWs would require small sliding to print NWs at a large scale. For these reasons, contact-printing is one of the most promising technique for LAE devices based on NWs.

The contact-printing technique has also been used for transferring and aligning NTs over large areas. The initial study in this area utilized a donor substrate consisting in vertically aligned NTs. By knocking down the NTs with a roller, horizontally aligned NTs can be obtained [283]. Further studies realize the transfer and alignment of NTs by pressing and sliding the donor substrate towards the receiver substrate surface (i.e. the contact-printing approach), using either vertically aligned [286] or randomly distributed NTs mesh on the donor [48]. Contact-printing of NTs has been carried out on various receiver substrates, including non-flat pre-patterned substrates and transmission electron microscopy (TEM) grids [48]. This approach enabled the direct printing of NTs onto substrates with pre-patterned electrodes, allowing the rapid and efficient fabrication of NT-based devices. The transfer yield of contact-printing process has been improved by using chemical treatments of the pre-patterned electrodes surface [287], which is promising for the fabrication of LAE devices.

Contact-printing of NTs has been demonstrated to exhibit different transfer mechanisms compared to NWs, mainly due to the as-grown structure of NTs [48]. Briefly, NTs synthesized on top of the donor substrate consist of two sublayers, comprising a bottom NTs layer, which interacts strongly with substrate, and a top randomly distributed NT-based network, which only holds a weak interaction with the bottom NT layer. During the contact-printing process, detachment of the bottom layer is almost negligible, as this occurs during the transfer process mainly in the top layer due to the far weaker interaction.

Stamp-printing. Another alternative dry-printing-based approach utilizes an elastomeric stamp to transfer nanostructures from the donor growth substrate to the receiver substrate. This approach is based on the stamp-printing technique [288]. Figure 3.4(a–c) shows a 3D schema of the transfer of CNTs using this technique. As mentioned, CNTs are typically fabricated by CVD mainly due to the excellent electrical properties of the resultant nanostructures [289, 290]. The high performance of CVD CNTs is due to the fact that they are fabricated with relatively long lengths (in the range of hundreds of microns) and are nearly free of bundles. In addition, CVD allows the growth

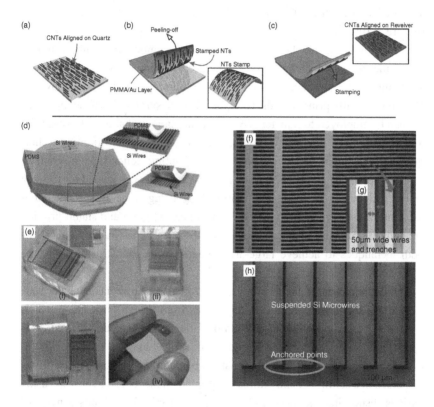

Figure 3.4 (a–c) 3D schematic illustration of a conventional stamp-printing experimental procedure used to transfer CNTs from a quartz substrate to a foreign substrate utilizing a polymeric stamp. (d) 3D schema of stamp-printing experimental procedure of Si NWs from SOI wafer to flexible substrates. (e) Images of Si wires stamp-printed on a pre-patterned circuit layout, exhibiting (i) wires picked up by PDMS stamp, (ii) PDMS stamped on adhesive SU8 layer, (iii) wires transferred to substrate after removing PDMS, and (iv) wires under bending conditions mode. (f,g) As-etched 50 μm Si wires on SOI. (h) Optical micrograph of the under-etched Si microwires, before their transfer to PDMS stamp. Reprinted with permission from Khan et al. [25] Copyright © 2016, IEEE.

of aligned CNTs based on electric field [291], laser ablation [292], gas flow [293] and the most straightforward method called *surface-oriented growth* [294]. These methods – and especially the latter – allow highly dense and well-aligned CNTs to be obtained on the substrate (Figure 3.4[a]). The transfer-printing of these CNTs is commonly carried out by firstly depositing a metallic layer (e.g. Au) on top of the CNTs, then, using a transfer substrate typically consisting of PDMS [295], forming a CNTs stamp (Figure 3.4[b]), and finally, the Au/CNTs layer being transferred to the receiver substrate (Figure 3.4[c]).

Contact-printing can be a complementary technique for stamp-printing, which means that contact printing can be used to print NWs from the growth substrate to a foreign receiver substrate, resulting in a highly aligned lawn of NWs horizontally printed on the receiver substrate surface. Then, stamp-printing can be employed to transfer NWs to an alternative substrate. However, the total transfer yield obtained by combining contact- and stamp-printing techniques could be lower than when using contact-printing alone. As in the contact-printing approach, the stamp-printing of NWs can also be scaled up to large areas, and this is dramatically improved by using large-area NWs donor substrates. In this regard, we can say that contact-printing and stamp-printing are both compatible with large-area donor substrates, which in fact would allow both techniques to achieve large-area integration of nanostructures such as NWs and NTs.

The combination of stamp-printing and contact-printing has been successfully probed, e.g. transferring Si microwires from silicon on insulator (SOI) wafers to both rigid and non-conventional flexible substrates [25, 35]. This approach has been carried out by means of: (a) wet-assisted process, and (b) dry-transfer printing. In the wet-assisted approach, Si microwires are firstly dispersed in a solution, being ultimately printed on the receiver substrate using various dispensing methods [25]. The resulting transfer exhibits a mesh of microwires randomly deposited along the receiver substrate surface, which makes it difficult to keep track of the desired finished or doped surfaces for further processing. In contrast, dry-transfer printing results in a deterministic assembly of oriented microwires on the receiver substrate. Under dry-transfer printing conditions, there can also be two approaches, comprising: (a) flip-over and (b) stamp-assisted transfer printing. The former consists of a single step-based process in which the adhesive surface of the receiver substrate is brought in conformal contact with the Si microwires and then pulled over to pick the microwires from the donor substrate surface. However, with this method the top surface of microwires (which could be doped for the development of devices after transfer is achieved) faces towards the transfer substrate, hindering the realization of metal contacts in post-processing steps. Given this drawback, stamp-assisted transfer printing (which involves two flip-over steps) is preferred. Figure 3.4(d) and (e) show a 3D schema and images, respectively, of the fabrication procedure followed to transfer Si microwires (Figure 3.4[f,h]) through a stamp-assisted transfer printing approach.

Roll-printing is an extension of the contact-printing approach. It is based on the direct contact and sliding of the NWs substrate along the receiver substrate surface, or vice versa, leading to the transfer of the well-aligned NWs along the sliding direction of the receiver substrate surface. The main distinction of roll-printing is the use of a cylinder (also known as a roller) to hold either the donor substrate, the receiver substrate or both [53, 296]. Figure 3.5(a) shows a 3D schematic illustration of the roll-printing configuration, presenting a flexible receiver substrate rolled along the donor substrate surface consisting of vertically aligned NWs. The pressure exerted by the cylinder on the donor substrate surface occurs only along the tangential line formed once both substrates are brought into contact, which allows accurate control over the areas with NWs that are subjected to the contact pressure. Roll-printing has been demonstrated to be compatible with non-conventional substrates (plastics, paper, textile, etc.) [53, 296] and highly scalable towards larger areas (even wafer-scale fabrication), and, due to its cylindrical geometry, is also compatible with R2R technology. Accordingly, roll-printing is a promising technique for fabricating LAE applications. The roll printing of NWs and NTs has scarcely been investigated and reported in the literature. So far, two main approaches have been successfully utilized to roll-print NWs, namely: i) differential roll-printing (Figure 3.5[b]) [53] and ii) roll transfer-printing (Figure 3.5[g]) [296].

The differential roll-printing method uses a roller as the growth substrate for NWs. By rolling the cylinder with NWs over the receiver substrate, the NWs can be transferred and aligned along the rolling direction. It should be noticed that the mismatch of the roller and wheel radii presented in the schematic of Figure 3.5(b) favours the transfer-yield of NWs with a density up to 9 NWs/μm. In contrast, the use of the same radius for both the roller and the wheel dramatically decreases the transfer yield, resulting in a low NWs density down to values around 0.04 NWs/μm. This phenomenon was explained as being due to the requirement of not only a direct contact between donor and receiver substrates but also a directional sliding motion along the tangential interface formed between the roller and flat substrate producing a shear force that bends, fractures and finally transfers/aligns the NWs on the receiver

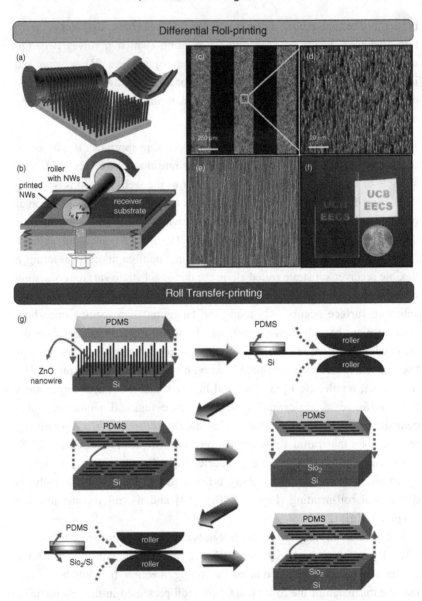

Figure 3.5 3D schematic illustrations of differential roll-printing procedure using either (a) a flat donor substrate with vertically aligned NWs or (b) a donor roller with NWs. Morphological characterization of roll-printed NWs carried out by both (c,d) optical microscopy and (e) SEM. (f) Successful roll-printing of NWs over large areas and on various substrates, including glass and paper. Reprinted with permission from Yerushalmi et al. [53] Copyright © 2007, AIP Publishing. (g) 3D schematic illustration of the

substrate. From the morphological characterization of the roll-printed NWs (Figure 3.5[c–e]), the transfer performance is not affected by the rolling velocity within a certain threshold (2 cm/min). A large printing velocity above that threshold leads to unstable printing, i.e. non-uniform NW alignment and poor coverage. The performance of the roll-printing method is similar to that obtained for conventional contact-printing (Figure 3.3), i.e. both methods have demonstrated the ability to fabricate 2D and 3D devices, similar NW densities, wafer-scale NW transfer and compatibility with both rigid and flexible substrates. In terms of the large-area transfer of NWs, the differential roll-printing approach shows a natural advantage over conventional contact-printing, where the latter requires a large donor substrate with a high density of uniform NWs, whereas the former has preliminarily demonstrated the ability to carry out sequential roll-printing processes to increase the NWs density over the receiver substrate surface [53].

Alternatively, the roll transfer-printing method combines roll-printing (Figure 3.5[a,b]) and stamp-printing (Figure 3.4) approaches to transfer and align NWs on foreign substrates [296]. This method is schematically described in Figure 3.5(g) and involves: 1) stamp-printing of a polymeric stamp atop the NWs substrate surface; 2) lamination of the resulting sample between two rollers; 3) peeling off of the stamp with well-aligned NWs; and 4) a second lamination/peeling off process to transfer NWs to the foreign substrate. Compared with differential roll-printing, the roll transfer-printing technique reaches faster printing velocities (20–80 cm/min), which are around one to two orders of magnitude faster than those possible with the differential roll-printing approach. However, the main drawback of the roll transfer-printing method is the utilization of several fabrication steps, as shown in Figure 3.5(g), that can drastically hinder the reproducibility and reliability of the process, especially for LAE applications.

Combing also known as *nano-combing*, resembles contact-printing over pre-patterned receiver substrates [37, 39]. Briefly, NWs are anchored to defined

---

Caption for Figure 3.5 (cont.)

experimental fabrication steps used for roll transfer-printing of NWs. This particular approach combines both roll- and stamp-printing mechanisms. Reprinted with permission from Chang et al. [296] Copyright © 2009, IOP Publishing Ltd.

areas of a substrate surface and then drawn out over chemically distinct regions of the substrate (Figure 3.6[a]). With this technique, there are two coexisting processes, also observed in the conventional contact-printing technique, which are the NW anchoring and the NW directional alignment. In contrast to contact printing, the combing approach allows the observation and therefore control of both processes individually. While the anchoring force is necessary during the whole NW printing process, for realizing the task of NW detachment from the donor substrate and reattachment to the receiver substrate, such a force dramatically hinders the effect of directional alignment [37, 39]. This conclusion has also been reached in contact-printing experiments, where high crossing defect density and the difficulty in realizing the precise registration and position of single NWs in predefined positions is mainly associated with the use of excessive anchoring forces [33]. The combing method has demonstrated an excellent potential to overcome the aforementioned drawbacks of contact printing, exhibiting the successful reduction of the crossing defect density down to 0.04 NWs/μm by tuning both anchoring and combing forces (Figure 3.6[a]). In terms of the realization of a single NW device (Figure 3.6[b]), the combing method shows great advantages over traditional contact-printing. By controlling the predefined anchor window, the success rate for the realization of a single NW device is ~90% (Figure 3.6[b]): significantly higher than the success rate (~60%) achieved using conventional contact-printing processes (Figure 3.3) [33]. It should also be noticed that while the combing method gives a higher control of NW alignment (~98.5% of the NWs aligned to within 1% of the combing direction) and registration, the resultant NW density, it is not comparable to the conventional contact-printing process (~9 NWs/μm) [33]. As a result of that, the combing method shows more promise in the realization of single NW-based devices (Figure 3.6[a]) and NW-crossed geometries (Figure 3.6[c]), where NW positioning and alignment are critical parameters for the development of LAE applications. On the other hand, the conventional contact-printing method could be used to achieve high-density NWs for LAE applications.

3.2 Integration by Assembly Techniques

The large-area integration of NWs and NTs on both conventional and non-conventional substrates has also been carried out through solution-based methods. These methods have attracted a lot of attention during recent decades for flexible LAE applications, mainly due to their compatibility with plastics, textiles and papers, as well as their cost-effective and easy-to-develop procedures. The standard procedure comprises the transfer of

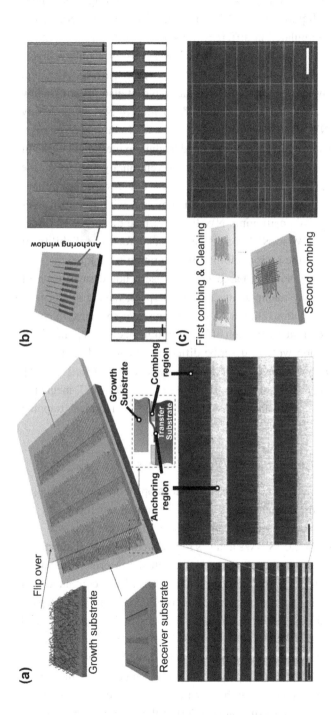

Figure 3.6 3D schematic illustration and SEM characterization of combing procedures used to carry out the assembly of (a) multi-NW arrays patterns with a printing area width limited by the combing region width (scale bars: left-SEM 50 μm, right-SEM 10 μm), (b) single-NWs (scale bar 2 μm), and (c) orthogonally aligned NWs (scale bar 1 μm). Reprinted with permission from Yao et al. [37] Copyright © 2013 Macmillan Publishers Ltd.

NWs from the growth substrate to an organic solvent, resulting in a suspension of NWs. In this liquid medium, NWs present more motion freedom, and therefore can be manipulated in different ways to trap and to align them on either flexible or rigid substrates. One of the lowest-cost procedures, involving NWs assembly in solution, is well known as *drop-casting*. Although this technique is low-cost, easy to develop and rapid, it also presents a low assembling yield mainly due to the random deposition of NWs without any control over both NW linear/surface density and NW directional alignment, hindering its applicability to LAE devices and leading to inefficient waste of material [297]. In this regard, drop-casting is conventionally utilized for the developing of proof-of-concept – i.e. early stage – devices for preliminary studies, but is not suitable for flexible LAE devices, where control over parameters such as NW positioning, NW linear/surface density, NW directional alignment or NW-to-NW spacing are crucial for the reliability of electronic devices and the reproducibility of integration techniques.

As mentioned above, solution-based methods use a liquid suspension of NWs to allow their easy manipulation in the 3D space, having an accurate control over their orientation, alignment, density and inter-spacing, which are essential parameters for the fabrication of high-performance LAE applications. In the following section we will describe different techniques based on this method, including bubble-blown [36], Langmuir–Blodgett [59, 186, 298–301], flow-assisted alignment [91, 302], electric-field [55, 91, 303–307], magnetic-field [57, 308–310], electrostatic-assisted techniques, and optical trapping [311–314].

3.2.1 Fluid-Assisted Techniques

Bubble-Blown (BB) technique was originally developed for manufacturing of large-area plastic films, which makes this technique an excellent tool for developing LAE devices based on nanostructures. BB technique was first used to transfer Si NWs, CdS NWs and CNTs to different kinds of substrates [36]. In BB, a molten polymer is extruded and inflated in order to obtain a balloon shape, which can be collapsed and slit to form large continuous flat films, exceeding 1 m in length at high rates of 500 kg/h [315]. The basic steps of BB consist of: (i) the preparation of a homogeneous, stable and controlled polymer suspension of NWs/NTs; (ii) the transfer of the suspension uniformly on top of a die which is blown into a single bubble using N_2 flow; (iii) finally, the transfer of NWs/NTs from the polymer to the substrate during the expansion of the bubble by direct contact between the bubble and the substrate as represented in the schematics of Figure 3.7(a). NWs are represented as red

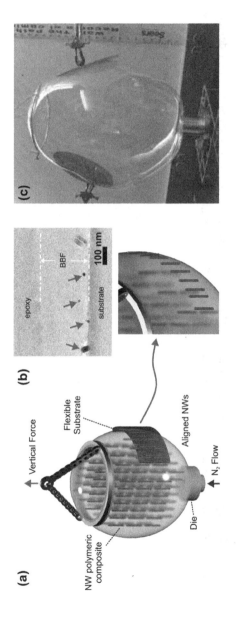

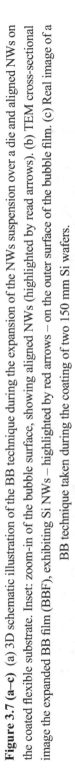

Figure 3.7 (a–c) (a) 3D schematic illustration of the BB technique during the expansion of the NWs suspension over a die and aligned NWs on the coated flexible substrate. Inset: zoom-in of the bubble surface, showing aligned NWs (highlighted by read arrows). (b) TEM cross-sectional image the expanded BB film (BBF), exhibiting Si NWs – highlighted by red arrows – on the outer surface of the bubble film. (c) Real image of a BB technique taken during the coating of two 150 mm Si wafers.

Figure 3.7 (cont.) (d–f) (d) BB coating on a 125 mm Si wafer. Inset: dark-field images presenting well-aligned Si NWs at different locations along the wafer. (e) Image of Si NWs BB-transferred to a 225×300 mm^2 plastic substrate; inset: histogram of the angular distribution of above 400 NWs analyzed in regions 1–6 marked in the image. (f) Dark-field optical image of a top-gated Si NW-FET device; inset: optical image of a 4×4 Si NW sub-array [36]. Reprinted with permission from Yu et al.[36] Copyright © 2007, Springer Nature.

cylinders inside a translucent polymer, where some of them are already aligned along the flexible substrate surface. As is also observed in this figure, the vertical expansion of the bubble is assisted by a vertically moving ring that keeps the bubble centred over the die. NWs passing through the die's circular aperture are subjected to a shear stress that produces their alignment along the direction of principal strain, i.e. the direction of the flow [58, 316]. During the BB process, the pressure difference between the inner and outer walls of the bubble tends to cause aligned NWs to drift to the outer surface of the bubble (see inset of Figure 3.7[a]). This was observed by TEM as shown in Figure 3.7 (b), where NW locations are highlighted using red arrows [36].This result makes the transfer procedure more effective, since there is a direct contact between the NWs and the receiver substrate.

Figure 3.7(c) shows a real image of the Si NW-based bubble expansion, and the coating of two 150 mm Si wafers at the same time. High-magnification dark-field images obtained at different points of the coated wafer, presenting individual Si NWs well-aligned along the upward expansion direction of the bubble (Figure 3.7[d]). The good alignment of the NWs in the bubble allows transferring NWs with lengths between 10 and 15 µm and slight deviations with respect to the flow direction lower than 10°. Optical analysis of the BB-transferred Si NWs shows centre-to-centre spacing and NW density of 3–50 and $4\text{-}400 \times 10^4$ cm^{-2}, respectively, are observed to be strongly dependent on the NW concentration in the suspension.

The BB technique has also been successfully utilized to align SWNTs and MWNTs with lengths of 1–2 and 20–25 µm, respectively, onto different substrates, including both 75–200 mm Si wafers and flexible substrates [36]. In the case of Si wafers, transferred SWNTs exhibit an average separation of 1.5 µm, and 90% of them are aligned to within 5° of the average orientation.

The main advantage of BB with respect to the rest of the available techniques to transfer nanostructures such as NWs and NTs to foreign substrates, is its compatibility with almost any kind of substrate and its great scalability towards large areas. In this regard, the area of the substrate that can be covered using BB is mainly limited by the diameter of the bubble that can be obtained during its expansion. This diameter can be further increased by increasing the diameter of the die and having a greater control of the expansion procedure (N_2 flow, vertical force magnitude, etc.). Since this procedure allows bubbles to be obtained in polymers with diameters of 1–2 m, BB is therefore an interesting technique to carry out the transfer of NWs to large-area substrates. Accordingly, SWNTs and Si NWs are BB-transferred to 200 mm Si wafers and 225×300 mm^2 plastic sheets (Figure 3.7[e]), respectively, exhibiting remarkable uniformity and good alignment along the primary expansion direction. Other substrates, including highly curved surfaces such as half-cylinders and open frames were also successfully use for BB-transferring of both Si NWs and NTs [36].

The preparation of the polymeric NWs suspension for BB assembly procedure is considered one of the main challenges impeding the successful rate of the assembly procedure. This preparation involves a sequence of surface chemical treatments of the NW (functionalization) prior to the NWs transfer to the polymer. The functionalization of the NW surface would enable the optimization of the suspension uniformity, preventing the formation of NW aggregations. These aggregations can reduce the dynamic range of NW weights that can be used to control parameters such as the resultant NW spacing and density. The optimization of the functionalization procedure is therefore an important pillar of this technique.

To summarize, BB is a promising technique for LAE mainly due to advantages including a high degree of alignment, controlled density of NWs and NTs, large-area coverage ability, and capability of transferring to a wide range of substrates. As an example of the great potential of this technique, BB has been used to fabricate arrays of independently addressable NW-based FETs. Figure 3.7(f) shows a 4 × 4 FET sub-array belonging to a 3 × 3 repeating transistor array, where each of the elements of the overall array consists of 400 independently addressable multi-NW transistors in a 20 × 20 array (not shown in the figure). The resultant performance of the BB technique allows rough control of the number of Si NWs per device, through the wt% in the polymeric bubble, with 12 Si NWs per device as shown in Figure 3.7(f).

Langmuir–Blodgett (LB) technique is another solution-based approach, promising for hierarchically organizing high-aspect ratio nanostructures such as NWs and NTs as building blocks over large areas [300]. In LB, NWs are not soluble in liquid but are floating on the liquid surface like 'logs on a river' (Figure 3.8[a]). This scenario is observed in cases where NW density is low or can also be achieved by functionalizing the NW surface with self-assembled monolayers, e.g. making the NW surface hydrophobic, allowing them to float on the medium of water.

LB technique is commonly carried out following two different procedures, called here: 1) stamping-LB and 2) Langmuir–Schaefer-based LB, which are schematically described in Figure 3.8(a–c) and Figure 3.8(d–f), respectively. The main difference between both procedures is the way through which NWs are transferred to the substrate. On the one hand, stamping-LB uses two barriers to exert a compression force on the NWs, leading to their alignment parallel to

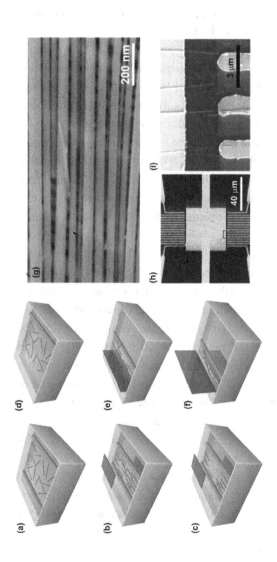

Figure 3.8 3D schematics of (a–c) stamping-LB and (d–f) Langmuir–Schaefer-based LB. Firstly, functionalized NWs are spread over the liquid solvent (a,d). Then, the barrier (e) or pair of barriers (b) compress the solution volume, leading to the alignment of NWs orthogonally to the compression direction, also increasing the NWs density. Finally, the receiver substrate is (c) stamped directly on the NWs solution, peeling some of the aligned NWs, or (f) dipped into the solution solvent, and slowly pulled out at a controlled vertical speed while the barrier moves at a specific horizontal speed. (d–f) In the latter, the synchronization of both speeds (i.e. barrier and receiver substrate speeds) allows the NW assembly onto the substrate surface, resulting in an array of aligned NWs. (g–i) SEM images of Si/SiO$_2$ NWs assembled by LB. Reprinted with permission from Whang et al. [317] Copyright © 2003, American Chemical Society.

the barriers (Figure 3.8[b]); then, substrate is horizontally brought into direct contact with the liquid surface, resulting in aligned NWs transferred to the substrate (Figure 3.8[c]). On the other hand, the Langmuir–Schaefer-based LB approach utilizes only a barrier to compress the NWs against the reservoir wall (Figure 3.8[d]), while the receiver substrate – dipped in the NWs solution – is slowly pulled out of the NWs solution (Figure 3.8[e–f]) [300]. The Langmuir–Schaefer method is strongly influenced by the speed of both the movement of the barrier and the extraction of the substrate.

The chemical surface properties of the NWs play an important role during the compression step described for the above LB approaches. As mentioned, the surface functionalization of the NWs not only makes NWs hydrophobic, enabling NWs to float on the solution, but also prevents the aggregation of NWs. The functionalization process depends on the surface properties of each kind of NW. For example, both Ag and Si NW surfaces can be rendered hydrophobic after functionalizing with 1-hexanodecanethiol [186], and 1-octa-decylamine [317], respectively. In this regard, LB requires a thorough knowledge of the chemical and electrical properties of the NWs surface in order to carry out an effective functionalization.

The space available in the reservoir has also a strong influence on both the alignment and the resulting surface density of NWs (NWs per area units) depending on the space available in the reservoir. Since the NWs are floating over the liquid solution, the reduction of this space enhances NW alignment and increases their surface density by compression effects. Typical surface pressure is in the range of 50–60 mN/m [300]. Parallel NWs are transferred to the receiver substrate in a layer-by-layer process resembling a 2D ordering of liquid crystal. Figure 3.8(g) shows a mono-layer of Ag NWs LB-aligned onto a Si wafer. The high packing density observed in that SEM image confirms the advantage of this technique for organizing these building blocks into functional nanostructures with unprecedented compactness. In the particular case of metallic NWs, the coating of the substrate surface with a high density of these NWs could lead to strong enhancement, for example of the Raman spectroscopy intensity, being therefore very attractive for the development of highly sensitive and selective biosensors [186]. In the case of Si NWs LB-transferred onto planar substrates, centre-to-centre spacing down to 0.2 µm has been obtained without observing aggregation (Figure 3.8[h–i]) [317]. This is so far one order of magnitude lower spacing than that obtained in the BB approach for assembling Si NWs [36].

However, in LB experiments a further reduction of the NW spacing below 200 nm leads to the formation of aggregations due to the strong inter-NW attractive forces. In order to overcome this issue, the NW centre-to-centre spacing obtained after a LB process has been further reduced by coating NWs with a thin sacrificial layer that can be removed after the LB process [318]. Briefly, Si-SiO$_2$ core–shell NWs (50 nm Si core and 20 nm thick SiO$_2$ shell) are firstly aligned using LB and transferred onto a Si planar substrate (Figure 3.9[a]). The SEM image shown in Figure 3.9(b) evidences the high packing density of the LB-transferred Si-SiO$_2$ NWs. Following the transfer, selective anisotropic etching is carried out by reactive-ion etching (RIE) to remove only the SiO$_2$ shell of the NWs (Figure 3.9[c]). The SEM image after etching (Figure 3.9[d]) shows an increase of the centre-to-centre spacing up to 40 nm, which is so far two orders of magnitude lower than the spacing obtained by BB for the same kind of NWs [318].

Taking advantage of close-packed parallel NWs resulting from the LB-transfer, NWs can be used as a mask for the deposition of metal nano-lines [318]. Figure 3.9(e) shows the resultant thermal evaporation of Cr thin film (15 nm thick) on top of the NW mask; then NW shadow mask is removed, resulting in well-defined parallel metal lines with a separation of 40 nm (Figure 3.9[f–h]). Since the line spacing is only dependent on the NW diameter, Figure 3.9[g,h] present SEM images of Cr lines with different spacings of 0.6 and 0.3 μm. This technique, therefore, offers excellent results comparable to the state-of-the-art techniques such as extreme UV lithography [38]. In this regard, LB allows the assembly of NWs to be carried out over large areas (20 cm^2), exceeding most other lithography methods, and can be extended even further to cover much larger areas using modified LB systems [319].

LB can also be utilized to fabricate more complex structures including patterned parallel and crossed NW arrays through LB hierarchical processes [317, 318]. Lithography is used to define patterns into repeating parallel Si NWs (Figure 3.9[i] – left) and Cr lines (Figure 3.9[i] – centre) arrays of controlled dimensions. In this regard, arrays of crossed Si NWs are also patterned over large areas, as shown in the right panel of Figure 3.9(i). Nano-electronic devices can be fabricated from these hierarchical NW arrays by using complementary techniques such as focused ion beam (FIB) to define series of parallel finger electrodes, contacting NWs in each of the arrays. While success has been achieved with LB for assembling single-layer arrays of functional NW devices based on NWs with diameters above 20 nm, this approach has exhibited less success yield in assembling smaller NWs (i.e. NWs with diameters below 15 nm).

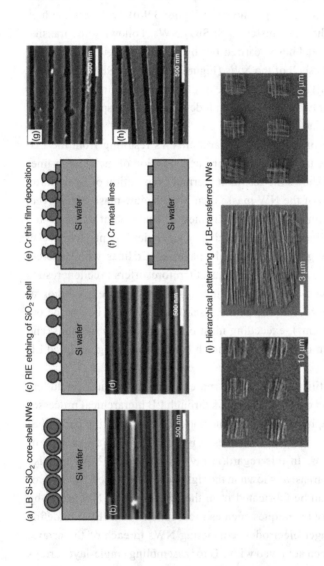

Figure 3.9 (a) 2D schematics and (b) SEM image of Si-SiO₂ NWs LB-transferred on Si wafer. (c,d) RIE of the SiO₂ shell. (e) Deposition of 15 nm Cr layer using the NW mask designed in (c,d). (e,f) Schematics and (g,h) SEM images of Cr lines after NW mask removal. Reprinted with permission from Whang et al. [318] Copyright © 2003, American Chemical Society. (i) SEM images of hierarchical patterning: (left) parallel Si NWs, (centre) parallel Cr lines, and (right) crossed Si NWs. Reprinted with permission from Whang et al. [317] Copyright © 2003, American Chemical Society.

Flow-Assisted Assembly's fundamental principle is based on fluidic mechanisms, comprising the use of micro-/nano-fluidic channels to align and to place nanostructures such as NWs and NTs at specific points over the receiver substrate surface [316]. In this scenario, the fluid motion exerts a shear force along with the boundaries of the NWs in order to minimize the drag forces, resulting in the alignment of the NWs along the flow direction. Flow-assisted alignment is compatible with flexible substrates and exhibits potential scalability for integrating NWs over large areas. This approach has demonstrated the successful alignment of different kinds of semiconductor NWs, including InP, Si and GaP NWs [316]. These works have shown the hierarchical assembly of NWs, controlling both the separation and the spatial location of the NWs, resulting in parallel and crossed arrays of NWs. Firstly, different NW suspensions were prepared, including GaP, InP and Si NWs suspended in ethanol solution. Then, solutions were passed through fluidic micro-channels formed between a polymeric film of PDMS and a planar substrate such as a Si wafer (Figure 3.10[a]). Morphological characterization of the resulting samples exhibits NWs aligned in the direction of fluidic flow (Figure 3.10[b]) [316]. In this approach, the effective area coated by aligned NWs is mainly limited by the micro-channel size, with the channel widths ranging from 50 to 500 nm and the lengths from 6 to 20 mm. In addition, both NW alignment and surface coverage were demonstrated to be controllable by flow-assisted parameters such as the flow rate and the duration. On the one hand, the degree of alignment can be controlled by the flow rate, observing a sub-stantial improvement of the NW alignment with the flow rate (Figure 3.10 [c]); this result can be explained within the framework of the shear flow [302], where higher flow rates lead to higher shear forces, resulting in a better NW alignment. On the other hand, the average NW coverage can be also controlled by the duration of the flow, observing a drastic increase of the assembled NW density with the flow duration, and showing densities around 250 NWs per 100 μm at a flow duration of 30 min, and reaching a NW-to-NW separation of around 0.4 μm (Figure 3.10[d]). This technique allows further reduction of the separation between NWs, showing values bellow 0.1 μm [316], which are much lower than those obtained by other fluid-assisted techniques such as LB or BB techniques [317].

One of the potential applications of flow-assisted technique is the possibility to fabricate crossed arrays of nanostructures following a layer-by-layer deposi-tion process. Figure 3.10(e) presents a 3D schematic illustration of this proce-dure, where a second flow-assisted procedure is performed on top of a substrate with aligned parallel NWs. A typical example of crossed NWs obtained

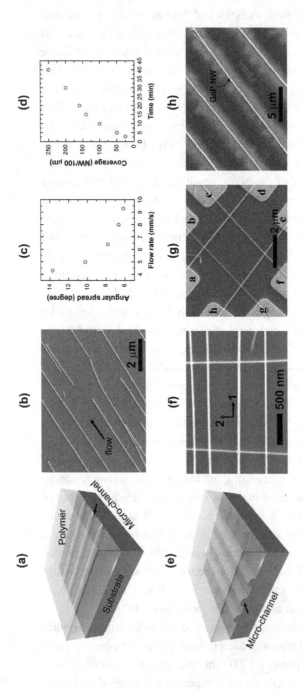

Figure 3.10 (a) 3D schema of the fluidic channel structures utilized for uniaxial flow-assisted alignment of NWs. (b) SEM image of the resultant aligned InP NWs on the SiO₂/Si substrate surface. (c) Angular spread of NWs with respect to the flow direction vs. flow rate. (d) Average NWs density vs time. (e) 3D schema of the fluidic channel structures using to fabricate crossed alignment of NWs. (f) SEM image of crossed arrays of InP NWs obtained by using a two-step-based fluidic-assisted assembly, comprising orthogonal flow directions for the sequential steps. (g) SEM image of a 2 × 2 cross array made by sequential assembly of n-type InP NWs by using orthogonal based fluidic assembly. (h) SEM image of parallel arrays of GaP NWs aligned on PMMA patterned surface with 5- and 2-μm separation. Reprinted with permission from Huang et al. [316]

Copyright © 2001, The American Association for the Advancement of Science.

through a flow-assisted technique is shown in Figure 3.10(f), where (1) and (2) represent the two sequential procedures used to obtain the resultant structure. Accordingly, the angle formed between crossed arrays of crossed NWs can be controlled by the orientation of the micro-channel [316]. The important feature of this layer-by-layer NW alignment method is that each alignment step is independent from the others, allowing the fabrication of both crossed homo- and hetero-junction-based NW structures, by simply changing the composition of the NWs solution at each specific step. Taking advantage of this function- ality, it is possible to fabricate crossed and monolithic architectures [316].

Flow-assisted assembly efficiency can be further improved by functionaliz- ing receiver substrate, making its surface more attractive for the NWs. Accordingly, NWs of different types, including GaP, InP and Si NWs, have shown better alignment efficiency on positively charged surfaces, e.g. NH_2- terminated monolayers, than on methyl-terminated monolayers or bare SiO_2 surfaces. Complementarily, chemically patterned substrates have been demon- strated to positively contribute to the organization of regular superstructures based on NWs. Under this approach, NWs tend to be preferentially assembled at specific positions of the substrate defined by the chemical pattern. However, it should be noticed that the use of patterned substrates is not enough for the proper alignment of NWs over large areas, resulting in NWs bridging and looping in the patterned areas [320]. In this regard, the use of a flow to assist the assembly of NWs along patterned substrate is a promising solution to overcome these issues.

3.2.2 Non-Fluid Assisted Technique

Electrical Field-Assisted Assembly. One of the essential features required for realizing a functional nanostructure-based device is the successful fabrication of a stable electrical connection between nanostructures and metallic electro- des. To achieve LAE applications based on NWs (or NTs), it is important to notice that the electrical contact stability and quality need to be realized at the nanoscale but over a macroscopic circuit layout covering a large area substrate. In this regard, the assembly of nanostructures such as NWs and NTs between metallic electrodes using electric fields has demonstrated an excellent potential to create high-quality nanoscale contacts, bridging pairs of electrodes through well-aligned NWs and NTs [55, 90, 91]. A non-uniform electric fields assembly technique, namely dielectrophoresis (DEP), has been used to trap and to align metallic NWs (Au [321], Rh [44], Ag [322]), semiconductor NWs (Si [44, 323], Se [318], InP [324], ZnO [55, 91, 306], CdSe [325], CuO [90], GaAs [307]), CNTs [305] and polymer nanofibers [326] between conductive electrodes for different applications such as photodetectors (Figure 3.11[a–d]) [55, 307], gas sensors [327], transistors [328], interconnects [329] and CMOS circuits [330]. In addition to single NW manipulation, DEP has been successfully used to assemble crossed NWs through a two-step process, applying electric fields in orthogonal directions [324].

DEP has demonstrated an increasing use for the controlled assembly of nanostructures at specific points along almost any kind of receiver substrate, which comprises not only control over the number of assembled nanostructures – through DEP parameters (Figure 3.11[e]) and DEP time (Figure 3.11[f]) – but also their alignment and location along the sample surface. DEP relies on alternating-current (AC) electric fields to align structures in solution at specific sites over large areas, at room temperature, and without the need of expensive tools. In DEP, the magnitude of either attractive (positive DEP) or repulsive (negative DEP) force depends on the polarizability of the NW, material dielectric constant and NW dimensions as described by the Clausius–Mossotti expression [304, 331, 332]. However, the lack of monitoring techniques to observe *in situ* a DEP assembly of nanostructures in a fluid, such as high-resolution microscopy in fluids, hinders the understanding of the DEP mechanisms that govern the alignment, trapping and overall assembly of nanostructures. Thus, new theoretical models are constantly developed to explain the assembly of nanostructures between electrodes through DEP [303, 304, 333, 334].

The working principle of DEP is based on local dipole moments induced along the nanostructure by the non-uniform electric field. In this regard, the aforementioned magnitude of the DEP force can be approximately calculated by the effective dipole moment (EDM) of the particle as demonstrated in CNTs [303]. However, EDM approximation is only valid for nanostructures with a size much smaller than the characteristic length of the electric field and is not valid for nanostructures comparable in size or larger than the gap between the metallic electrodes, which hinders the use of EDM to simulate any practical DEP experiments. Alternatively, multipolar correction terms have been added to the EDM model [303], enabling the simulation of larger particles. However, as in the case of the imaging method [334], both models are only applicable to simple problems of multiple dipole interactions of one or two spherical particles, hindering their use for DEP of NWs and NTs. To simulate the DEP force on high-aspect ratio nanostructures such as NWs and NTs in the vicinity of metallic electrodes with micro-sized gaps and under the effect of a non-uniform electric field, it is essential to solve the electric field as a boundary value problem and numerically compute the DEP force using the Maxwell stress tensor (MST) method. One of the most promising explanations about the DEP

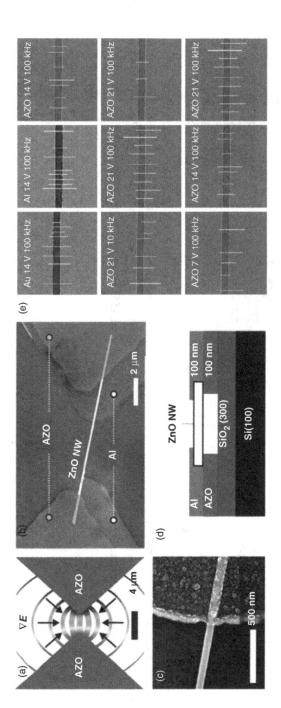

Figure 3.11 (a–e) (a) Electric field lines (line thickness and colour represent |E| and electric potential, respectively) and ∇E arrows for Al doped ZnO (AZO) electrodes. (b) SEM image of a single ZnO NW-based device whose edges are covered with Al (c) as detailed in the schema (d). (e) SEM images of ZnO NWs DEP assembled between Au, Al and AZO electrodes at different DEP conditions (i.e. amplitudes and frequencies of the alternating-current (AC) signal applied between electrodes). Reprinted with permission from García Núñez et al. [55] Copyright © 2013, IOP Publishing Ltd.

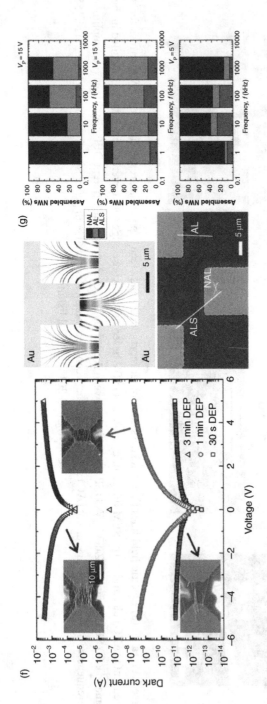

Figure 3.11 (cont.) (f–g) (f) UV photodetectors based on ZnO multi-NWs DEP assembled between electrodes for different times. Reprinted with permission from García Marín et al. [91] Copyright © 2015, IOP Publishing Ltd. (g) Analysis of the NW DEP alignment vs. electrode geometry. Reprinted with permission from García Núñez et al. [55] Copyright © 2013, IOP Publishing Ltd.

assembly of NWs has been described through a three-dimensional dynamic model based on the immersed finite elements method (IEFEM) and electro-kinetic theory [332].

With respect to LAE applications, DEP and other techniques, such as magnetic field assisted and optical trapping techniques, are well-known for their high-precision manipulation of micro-/nano-structures at low scales. Among them, DEP is the only technique that has demonstrated its scalability towards macroscales so far. In this regard, Se NWs have been DEP assembled into macroscopic fibres between electrodes and over large distances (above 5 cm), which are around three orders of magnitudes larger than the common assembly distances in DEP [322].

Electrostatic-Assisted Assembly. The high surface-to-volume ratio and aspect ratio of filamentary nanostructures such as NWs and NTs make them chemically attractive for interactions, such as chemical bonding, Van der Waals interactions and electrostatic interactions. These fundamental interactions have been used to assemble nanostructures such as V_2O_5 (Figure 3.12[a,b]) [335], CNTs (Figure 3.12[c–e]) [34], ZnO [336], AuNi [309], Au/Ni/Au [337], and Au/Pd/Au [338] NWs at selective places along chemically patterned receiver substrates such as Si, SiO_2, glass, Au and Al. The potential of electrostatic-assisted assembly allows the creation of micro-sized patterns consisting of either areas with well-aligned NWs or empty areas. For example, V_2O_5 NWs with a negatively charged surface have been selectively integrated on specific areas on a receiver substrate pre-patterned by positively and neutrally charged areas (Figure 3.12[a,b]). In this regard, V_2O_5 NWs show preferential assembly on positively charged areas induced by the electrostatic forces [335]. Another example of functionalization of the receiver substrate surface, to assemble nanostructures in selective areas along the substrate, is shown in Figure 3.12 (c–e) [34]. In this case, CNTs were assembled in specific areas using polar and non-polar molecular patterns on a gold layer. This approach was also used to carry out the preferential assembly of NWs, bridging pairs of Au electrodes, using DNA hybridization [338].

Magnetic Field-Assisted Assembly. In addition to electric fields and electrostatic forces, magnetic fields have also been utilized to assemble magnetic NWs (Ni[308]) and non-magnetic NWs (Au[57], Bi[57]) in solution between electrodes for different applications. Conventionally, ferromagnetic electrodes of micro-/nano-scale magnets are chosen, aiming to improve the resulting uniformity and precise control of the NW positioning over the receiver substrate surface. Following this strategy, Ni NWs have been successfully assembled between Co nanomagnets,

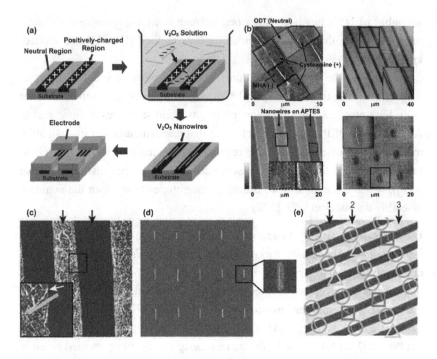

Figure 3.12 (a) 3D schema and (b) SEM images of surface-programmed assembly of V_2O_5 NWs. Reprinted with permission from Myung et al. [335] Copyright © 2005, WILEY-VCH Verlag GmbH & Co. KGaA, Weinheim. (c) Morphological characterization of CNTs near the boundary (white arrow, inset) between polar (left arrow) and non-polar (right arrow) molecular patterns on gold. Yellow bar represents a tangent to a bent CNT, showing the extent of bending due to lateral directional force. (d) Topography of an array of individual CNTs covering about 1 cm^2 of Au surface. Inset: single CNT. (e) Topography of an array of junctions with no CNTs (triangles), one CNT (circles) or two CNTs (squares), covering an area of about 1 cm^2. Arrows 1, 2 and 3 indicate octadecyltrichlorosilane (used to passivate the SiO_2 surface), 2-mercaptoimidazole on gold, and ODT on gold, respectively. Reprinted with permission from Rao et al.[34] Copyright © 2003, Nature Publishing Group.

using low-magnitude magnetic fields around 10 Oe (Figure 3.13[a–d]). The results exhibit a high trapping yield of 100% and high level of control over the assembling direction, i.e. the external magnetic field was successfully used to create NW-based patterns aligned along (Figure 3.13[c]) or orthogonal to the magnetic field direction (Figure 3.13[d]) [308].

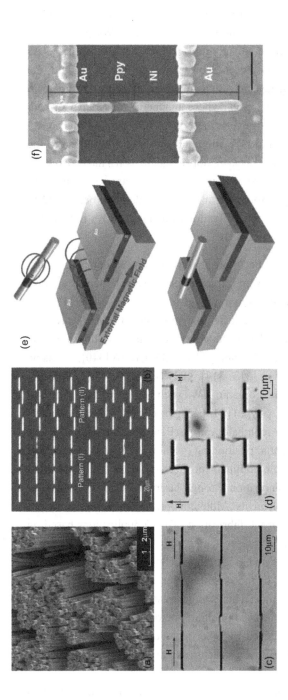

Figure 3.13 (a) SEM image Ni NWs with diameters of around 200 nm. (b) Optical microscope image of the template with nanomagnets (width between 100–400 nm and thickness of 100 nm). Optic micrograph of self-assembly Ni NWs arrays when an external magnetic field (10Oe) was applied (c) parallel and (d) orthogonal to the magnetization direction of the magnet template. Reprinted with permission from Liu et al. [308] Copyright © 2007, AIP Publishing. (e) 3D schema of the magnetic based assembly of multi-segmented metallic NW. The lines represent the magnetic field of the NW and the ferromagnetic electrodes. Bottom figure exhibiting the placement of different segments of the nanowire on the electrodes. (f) SEM image of single NW-based device fabricated by magnetic field-assisted assembly on pre-patterned ferromagnetic electrodes (scale bar = 1 μm). Reprinted with permission from Bangar et al. [57] Copyright © 2009, WILEY-VCH Verlag GmbH & Co. KGaA, Weinheim.

Alternatively, non-magnetic NWs can also be assembled between electrodes via magnetic fields, by coating the non-magnetic material with a magnetic material, or synthesizing multi-segment-based NWs (e.g. Ni/Au/Ni, Ni/Bi/Ni, gold–polypyrrole–nickel–gold) (Figure 3.13[e,f]) [57]. As shown in Figure 3.13(e), in the presence of an external magnetic field, the ferromagnetic electrodes behave as micro-magnets, creating strong local fields at the edge of the electrodes where magnetic poles are formed. These localized magnetic fields govern the NW-to-NW dipole-based interactions, leading to a preferential NW alignment in those areas close to the ferromagnetic electrode. In this scenario, non-magnetic NWs have also been demonstrated to be aligned between ferromagnetic electrodes using magnetic fields (Figure 3.13[e]). Moreover, magnetic fields have been used to spatially organize cells through the utilization of magnetic NWs in conjunction with patterned micro-magnets [310].

Optical Trapping Technique. For more than two decades, optical trapping has been successfully utilized to trap, transfer and assemble nanostructures, including NWs and NTs in a liquid environment, on top of the surface of a wide variety of rigid and non-conventional flexible substrates. In this regard, this optical trapping technique has demonstrated its great applicability to a wide range of dielectric nanomaterials, including CNTs [339], DNA [340], inorganic NWs such as Si [312, 313], Ag [313, 314], GaN (Figure 3.14) [313], SnO_2 [313], CuO [56], CdS [341] and organic NWs such as organic single-crystal copper phthalocyanine (CuPc, p-type) [342], single cells [343], single proteins [344] and copper hexadecafluorophthalocyanine (F16CuPc, n-type) [342]. The results show a nanometric spatial positioning (<1 nm) [345], and an excellent potential to control the particle confinement mechanism by tuning the light intensity, wavelength and polarization through lasers, acousto-optic modulators and holographic optical elements. Although optical-assisted assembly has not shown its full potential for the fabrication of LAE applications based on NWs (or NTs), this technique presents one main advantage compared to magnetic and electric field-assisted assembling techniques, which is the possibility to achieve better control on the assembly of individual nanostructures in specific places along the substrate. This is an interesting and powerful feature for developing e.g. single NW-based photonic, optoelectronic and electronic devices which cannot be achieved by other techniques due to their lack of selectivity in terms of number of nanostructures under the influence of electric and magnetic fields.

The working principle of optical-assisted assembly is similar to that described for DEP (Figure 3.11). Particles much larger and smaller in size

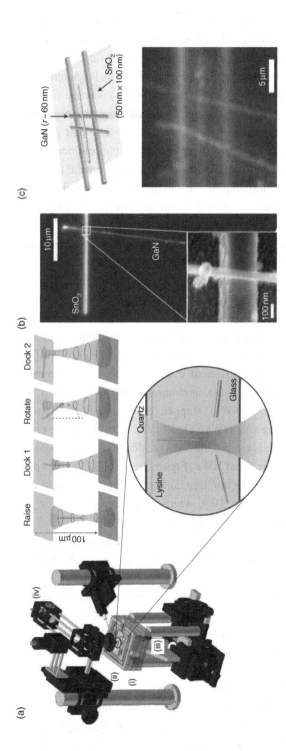

Figure 3.14 (a) Experimental setup used to carry out optical trapping of NWs, and 3D schematic illustrations, showing the four step-based NW positioning procedure. (b) Optical microscope image of a GaN NW laser-fused to a SnO$_2$ nanoribbon during an optical trapping process. Inset: SEM image of fused NW-to-NW junction. (c) 3D schema (top) and optical microscope image (bottom) of a crossed structure consisting in a layer-by-layer assembly of GaN NWs orthogonally aligned on top of SnO$_2$ nanoribbons. Reprinted with permission from Pauzauskie et al. [313] Copyright © 2006, Springer Nature

than the light wavelength can be theoretically modelled by conventional frameworks of the Mie limit [311] and the Rayleigh limit [346], respectively. In case of high-aspect ratio NWs (or NTs), the dimensions corresponding to the cross-section are typically below 100 nm, i.e. under the Rayleigh limit, whereas the length of a NW (typically in the range of μm) is under the Mie limit. In addition to the discrepancy between axial and longitudinal dimensions, NWs can absorb a relatively small amount of light and their polarizability is not well determined, hindering the modelling of the confinement by the optical potential. In spite of the modelling difficulties of optical trapping effects of NWs, the successful manipulation of single NWs through optical trapping, also called optical tweezers (or optical tweezing), has been reported. Comparing the results obtained from the optical trapping of different semiconducting and metallic NWs, it can be noticed that the aspect ratio and the material of the NWs drastically affect the confinement of a single NW in an optical potential. For example, assuming a single-beam optical tweezer system (Figure 3.14[a]) and NW suspension confined in a 100 μm thick cavity, it has been experimentally discovered that the laser beam can trap NWs one-by-one from the bottom of the cavity and align them on the top surface of the cavity by lifting the focus distance of the laser (Figure 3.14[b]) [313]. However, in this scenario, different materials with different aspect ratios have shown different behaviours, including stably trapped with no oscillations (GaN, ZnO and Si NWs), trapped but showing oscillations (SnO_2 NWs) and no trapping (Ag NWs). One of the most attractive characteristics of optical trapping is the possibility to realize complex structures based on NWs, e.g. crossed arrays of NWs (Figure 3.14[c]) [313]. Indeed, it has been demonstrated that the effect of light during the trapping process can result in a local fusing of crossed NWs, which can be promising for the development of p–n junctions and Schottky barrier-based NW devices. This can be explained by the increase in the local temperature at the joint of the crossed NWs, which can overcome the melting point of the materials and lead to the local fusing of the NWs.

In terms of applications, optical trapping has demonstrated great versatility in fabricating functional prototype photonic devices [313], CNT-based transistors [339] and circuit elements such as inverters, transfer gates, NOR and NAND gates [342], as well as chemical, mechanical and optical stimulation of living cells [313].

Table 3.1 summarizes the performance obtained by printing (Section 3.1) and assembly techniques (3.2) for the large-area integration of micro- and nanostructures, including advantages and disadvantages of each approach, as well as applications. For the sake of comparison, the main features of

Table 3.1 Suitability of printing/integration techniques used to fabricate LAE based on micro-/nanostructures.

Techniques	Advantages	Disadvantages	Applications
Inkjet printing	• Wafer-scale printing • R2R compatibility • Low-cost tools	• Complex ink preparation • Poor alignment control • Low density of printed nanostructures	• Supercapacitors [64] • Organic transistors [65] • Flexible electronics [66] • PV cell [67]
Screen printing	• Wafer-scale printing	• Complex paste preparation • Expensive tools • Poor uniformity and alignment • Thick layers	• Humidity sensors [272] • Gas sensors [273] • Chemical sensors [274] • Stretchable electronics [275]
Gapping method	• Wafer-scale printing • High alignment control	• Poor control on nanostructures • Use of high voltage	• Optical polarizers [276] • UV photodetectors [279] • Energy storage device [280]
Contact printing	• Wafer-scale printing • High alignment control • High density of printed nanostructures • Sequential printing	• Require treatment of receiver substrate • Complex and high-precision setups	• Artificial skin [347] • Flexible UV photodetectors [99] • FET [54] • Tuneable photodetectors [348]

Table 3.1 (cont.)

Techniques	Advantages	Disadvantages	Applications
Stamp printing	• Large-area transfer of nanostructures • Rapid transfer method	• Generation of structural defects • Stamp residues • Poor control of single nanostructures • Require expensive tools	• FET [349] • Plasmonic sensors [350]
Roll printing	• R2R and R2P compatible • Hierarchical printing • Printing on rigid and flexible substrates	• Complex setup • High-precision system • High-quality donor • Poor control of single nanostructures	• Large-area flexible FET [53] • FET [351]
Combing	• Accurate and individual control of contact and detachment mechanisms • High precision over the location and alignment of transferred nanostructures	• Poor control over single nanostructures • Low density of assembled NWs • Complex and high-precision system • Use of pre-patterned substrates	• FET [352]

Technique	Advantages	Disadvantages	Applications
Bubble-Blown	• Metre-scale fabrication • Compatible with rigid and flexible substrates • Sequential transfer of nanostructures • Crossed structures and complex architectures	• Low density of assembled nanostructures • Lack of control over single nanostructures • Complex system and preparation of nanostructures paste • Poor control over the alignment and position of transferred nanostructures	• PV cells [353] • Photodetector [354] • Gas sensor [354]
Langmuir–Blodgett	• Record values of density of assembled nanostructures • Wafer scale fabrication	• Require advance knowledge of nanostructures surface chemical properties • Chemical treatment nanostructures • Preparation of nanostructures solution • Complex system	• Biosensor [355] • Sensing applications [59]
Fluid-assisted technique	• Compatible with sequential fabrication • Crossed-architectures • Record values of low nanostructure-based layer compactness	• Require micro-channels • Preparation of nanostructures solutions	• FET [356]

Table 3.1 (cont.)

Techniques	Advantages	Disadvantages	Applications
Electric-field assisted integration	• Manipulation of single and multiple nanostructures	• Complex solution preparation • Poor scalability	• Photodetectors [55, 307] • Gas sensors [327] • Transistors [328] • Interconnects [329] • CMOS circuits [330]
Electrostatic-assisted integration	• Selectivity to multiple materials • Sequential assembly • High alignment and positioning control of nanostructures	• Require treatments of both substrate and nanostructures	• Supercapacitor [357] • Gas sensor [358]
Magnetic field-assisted integration	• High control over alignment and position of nanostructures • High control over single nanostructures • Hierarchical assembly	• Require magnetic materials • Poor scalability	• Optoelectronic devices [359] • Supercapacitor [360]
Optical trapping	• High control over alignment and positioning of single nanostructures	• Low control over multiple nanostructures • Poor scalability • Require complex and expensive setup • Require accurate calibration of the system	• Photonic devices [313] • Transistor [339] • Circuit elements such as inverters, transfer gates, NOR and NAND gates [342] • Chemical, mechanical and optical stimulators of living cells [313]

	Advantages	Disadvantages	Applications
Nanoimprint lithography	• Wafer-scale fabrication • High resolution at nanoscale • Low time consuming • R2R and R2P compatible • Low temperature procedures • Rigid and non-conventional flexible substrates	• Complex and expensive systems • Low pressure uniformity • Poor reproducibility • Non-continuous printing	• Flexible PV cell [361] • Optical polarizers [362]
Hot embossing lithography	• Wafer-scale fabrication • Easy and low-cost fabrication steps • Potential to create periodic nanostructures	• Defects from demoulding	• Interdigitated electrodes [62] • Anti-reflection coating [363] • Micro-fluidic channels [364]
Laser interference lithography	• Compatible with sequential fabrication • Crossed and complex architectures • Periodic nanostructures with different shapes	• Complex optical setup and calibration	• Nano-mask fabrication [365]

integration techniques based on lithography, whose main characteristics will be presented in the following section, have also been included in Table 3.1.

3.3 Integration by Lithographic Techniques

3.3.1 Nanoimprint Lithography

Nanoimprint lithography (NIL) is a nanopatterning-based technique that has demonstrated the capability to define patterns at the nanoscale with a resolution limited by the light diffraction or beam scattering. This technique has attracted great attention during recent decades for the development of LAE mainly because of its low time-consumption and compatibility with R2R [60] and roll-to-plate (R2P) manufacturing procedures [61]. In this regard, NIL has been proposed as an attractive technique for the development of a wide range of applications, including electronics, optoelectronics, photovoltaics and photonics based on organic and inorganic nanomaterials [361, 366–368], as well as for the fabrication of micro- and nano-channels for various fluidic applications [369]. Moreover, the low-temperature procedures conventionally used in NIL have demonstrated great performance for flexible electronics, allowing the fabrication of the above applications in non-convectional substrates such as paper, textile, plastics, etc. (Figure 3.15[a]) [361, 370, 371], and also in conventional rigid substrates (Figure 3.15[b]). However, drawbacks such as low pressure uniformity along large areas, discontinuity of the printed patterns over large areas and lack of safe demoulding of the resulting nanopatterns have delayed the utilization of NIL in practical LAE applications [61]. To overcome the aforementioned drawbacks, intensive investigations in the field have been carried out over the last decade [371–374], making NIL more reliable, rapid and suitable for the continuous and uniform printing of nanomaterials over large areas and flexible substrates. For all these reasons, NIL has attracted great attention for LAE based on nanomaterials. For example, periodic grating nanostructures consisting of long-aspect-ratio epoxysilicone-based lines with a width of 300 nm and length in the range of millimetres (Figure 3.15[c,d]) have been successfully fabricated by NIL over large areas (flexible: 4 × 12 inches; rigid: 4 × 10.5 inches) and on both flexible (Figure 3.15[a]) and rigid substrates (Figure 3.15[b]). This fabrication process was carried out in both R2R and R2P configurations, resulting in the impressive photographs presented in Figure 3.15(a,b). NIL has been also utilized to fabricate micro- and nano-fluidic channels for different purposes (Figure 3.15[e]). Nano-channels with a width and length, 75 nm and 120 nm, respectively, have been fabricated by NIL on a PMMA layer (Figure 3.15[e]). The great control over the resulting dimensions of the nano-channels has allowed the use of this technology for biological

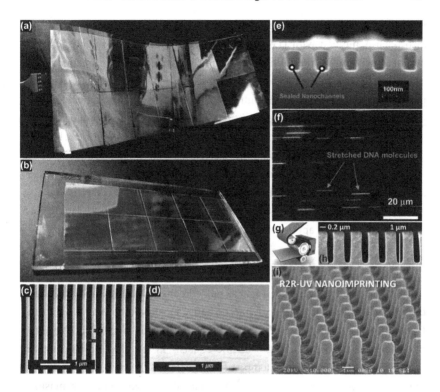

Figure 3.15 Photographs of epoxysilicone nanopatterns defined by NIL technique over large areas on (a) flexible and (b) rigid substrates. (c) Top-view and (d) cross-sectional SEM images of the nanopatterns. Reprinted with permission from Ahn et al. [61] Copyright © 2009, American Chemical Society. (e) Cross-sectional SEM image of nano-fluidic channels fabricated on PMMA by NIL. (f) Optical image of stretched DNA molecules assembled by nano-fluidic channels created by NIL. Reprinted with permission from Jay Guo et al. [369] Copyright © 2004, American Chemical Society. (g) 3D schema of R2R system developed for NIL fabrication processes. (h) Cross-sectional and (i) titled SEM views of patterned photoresist fabricated by NIL assisted by a UV curing process. Reprinted with permission from Leitgeb et al. [60] Copyright © 2016, American Chemical Society.

applications, comprising detecting, analysing and separating biomolecules. For example, taking advantage of the nanometric dimensions of the channels fabricated by NILs which are close to the persistence length of the DNA molecules, DNA was successfully stretching inside the nanofluidic channels (Figure 3.15[f]) [369]. In this regard, NIL can be considered a complementary technique for fluid-assisted techniques described later on. Recently, NIL has also been combined with UV curing processes using a UV-curable resistant system, speeding up the fabrication of uniform and continuous nanopatterns through R2R processes (Figure 3.15[g]) at speeds in the range of tens of metres per minute [60]. The accurate control over the viscosity, mechanical and surface properties of the above resistant has demonstrated high reproduction and fidelity of nanoscale features on rigid and flexible substrates (Figure 3.15[h,i]). This new approach, namely UV-NIL, has shown a significant reduction of the resulting defect concentration in the patterned resistant due to the water-solubility feature of the UV-imprint resistant. In addition, UV-NIL has also demonstrated high-throughput fabrication of metallic nano-grids (Figure 3.15 [h]), which has great applicability e.g. in photonics and optics.

3.3.2 Hot-Embossing Lithography

Hot-embossing lithography (HEL) has gained considerable attention for fabricating well-organized arrays of micro- and nanostructures over large areas through easy and low-cost fabrication steps [62]. HEL is a soft lithography-based method whose working principle consists in the use of a hard master stamp (Figure 3.16[a]) that is temporary contacted to the surface of a polymer (e.g. a resistant), creating a pattern that can be utilized either as masking resistant or transferred to a foreign substrate using stamp-printing techniques (see Section 3.1.2). HEL exploits the difference in the thermomechanical properties of the hard master stamp and the surface of a polymer (e.g. thermoplast). Essentially, the polymer is heated up to its viscous state, allowing it to be shaped under pressure by imprinting it with the hard master stamp. Once the polymer has conformed to the shape of the master stamp, the polymer is hardened by cooling and finally demoulded [62]. The latest advances achieved in fields such as on nanolithography and dry/wet etching procedures have benefited the fabrication of hard master stamps for the aforementioned HEL procedure, through the significant improvement of patterns with different shapes and nanoscale dimensions [375]. For example, HEL has been successfully used to fabricate large-area arrays of metallic nano-dots (diameter around 10 nm) with a nanoscale dot pitch of 40 nm [62]. The ability to emboss periodic nanostructures in polymeric materials, e.g. PMMA, permits the replication of various nanostructures such as dots, lines and meanders, with dimensions ranging from tens to hundreds of nanometres (Figure 3.16[b,c]). Such well-ordered nanostructures present great potential for the development of photonic structures and energy storage applications, where the uniformity of the pattern size and the periodicity of the nanostructures is crucial for the performance of

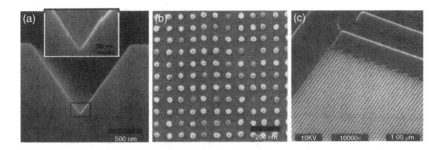

Figure 3.16 (a) SEM image of Si-based hard master stamp, featuring V-grooves with a sharp nanoscale shape (see inset). SEM images of (b) dots and (c) meanders replicated in PMMA by HEL. Reprinted with permission from Schift et al. [62] Copyright © 1999 Published by Elsevier B.V.

the application. HEL has also been used to fabricate interdigitated electrodes whose utilization in LAE is very well known, especially for sensing applications.

One of the key challenges of HEL, which is still under investigation, is the demoulding step. The understanding of the polymer behaviour at the nanoscale range makes the demoulding step critical to prevent the formation of defects in the resulting nanopattern. In this regard, polymer moulding with a low size of 10 nm has been successfully shown [376], which means that embossing hard master stamps of such a size is required. Although this achievement is promising for the definition of nanostructures in the range of tens of nanometres, their scalability to large areas needs to be further investigated.

3.3.3 Laser Interference Lithography

Laser interference lithography (LIL) is based on the interference of two coherent lights forming a horizontal standing wave along the surface of the sample, whose grating pattern can therefore be recorded e.g. on a photoresist (Figure 3.17[a]). The characteristics of the grating pattern depend on the wavelength, intensity and angles of the incident light. This interesting approach has been successfully used to fabricate periodic high-aspect ratio nanostructures with the shape of lines (Figure 3.17[b]) and nano-dots (Figure 3.17[c]) [63], the latter being possible due to the combination of multiple LIL steps carried out by rotating the sample through 90°. In contrast to the current drawbacks of the aforementioned NIL in fabricating large patterns consisting of nanostructures, LIL has demonstrated faster and larger scalability than NIL. In addition, LIL has shown great tuneability with respect to the shape of the resulting pattern,

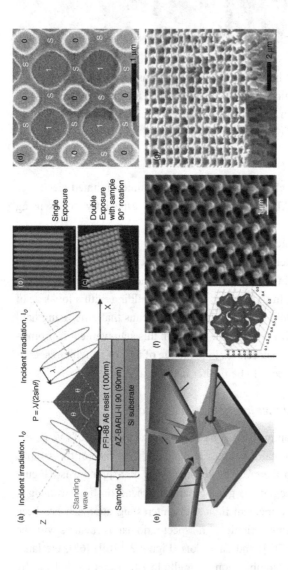

Figure 3.17 (a–g) (a) 2D schema of LIL fundamental principle, comprising two coherent lights incident on a resistant whose interference forms a standing wave. This wave records a periodic (b) line- or (c) dot-based pattern after a single- or double-exposure LIL process, respectively. (d) LIL parameters such as incident light intensity (d), wavelength and angle have been studied to produce complex patterns for different applications. Reprinted with permission from Xie et al. [377] Copyright © 2006, Elsevier B.V. (e) 3D schema of multi-beam-based LIL. (f,g) SEM images of SU-8 resistant-based 3D structure fabricated by LIL. Reprinted with permission from Jang et al. Copyright © 2007, John Wiley and Sons.

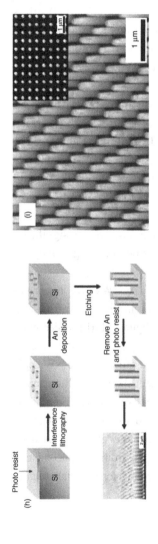

Figure 3.17 (cont.) (h–i) (h) 3D schema of the top-down synthesis of Si NWs using LIL. (i) SEM image of vertically aligned Si NWs synthesized by the process described in (h). Reprinted with permission from Choi et al. [378] Copyright © 2008, American Chemical Society.

allowing the formation of nanopattern with multiple shapes (Figure 3.17[d]). Although LIL is promising for the development of such complex patterns, further investigations are still needed to prevent negative effects of the multi-step LIL process, such as the overexposure of nanostructures pre-patterned in the neighbourhood of the pattern under fabrication [63]. In spite of this challenge, the possibility of fabricating nanostructures with different sizes and shapes along the same sample (Figure 3.17[d]) offers tremendous application opportunities for LIL in various LAE applications such as high-density data storage, micro-sieves for micro-filtration and sub-micron perforated membranes, templates for self-assembly and field-emission flat panel displays [63].

The functionality of the LIL technique can potentially be expanded by using a greater number of incident light beams (Figure 3.17[e]), which enables the creation of complex 3D networks and structures (Figure 3.17[f, g]). In this scenario, the theoretical simulation of the interference of multiple incident lights (inset of Figure 3.17[f]) is a matter of interest that would allow the creation of highly crystalline 3D structures [377]. The advances of LIL towards 3D integration of multiple layers based on nanomaterials is a promising direction for the development of LAE due to the significant increase of functional elements at the expense of device weight and size reduction.

LIL is also compatible with the top-down synthesis of NWs over large areas (Figure 3.17[h]) [378]. For example, Si NWs have been synthesized from a Si wafer through a top-down mechanism described in Figure 3.17(h), which combines LIL and catalytic etching (MACE). In this method, LIL is used to create a mask to form a metallic nano-mesh for the preferential growth of the NWs only underneath the metal areas (Figure 3.17[h]). The result shows a well-organized array of free-standing Si NWs on top of a Si wafer over large areas (> 1 cm^2). The accurate control of LIL over the distribution, shape and size of the nanopatterns makes this combined method promising for the creation of LAE based on semiconductor NWs. Since the synthesis is carried out from a semiconductor wafer, this technique can be expanded to other semiconductor materials such as III–V compounds.

4 Conclusions

In this Element we have reviewed the conventional and state-of-the-art technologies to integrate organic and inorganic filamentary nanostructures such as NWs and NTs over large areas for the development of high-performance rigid/flexible LAE applications. Integration techniques presented here involve the assembly of NWs/NTs synthesized by either top-down or bottom-up approaches. Both synthesis mechanisms and corresponding growth systems have been introduced and briefly described for each kind of nanomaterial, highlighting the resultant properties and advantages of NWs/NTs for different LAE applications. In this regard, integration techniques have been classified into printing and non-printing approaches. Comparing the figure of merit obtained by each technique, one can highlight: 1) LB, flow-assisted technique, contact-printing and combing for the accurate control of NW directional alignment, including crossed and monolithic architectures; 2) LB and contact-printing for high NW density; 3) LB, BB, contact-printing, screen-printing and inkjet-printing for large-area assembly (wafer-scale NW assembly); 4) optical trapping, electric field- and magnetic field-assisted techniques for precise control of single NWs/NTs over large areas. It is worth noticing that each technique mentioned above presents advantages and weaknesses for LAE applications. In this regard, the combination of multiple integration techniques, e.g. roll-printing + stamp-printing, dielectrophoresis + flow-assisted technique, or LB + stamp-printing, should be the most promising strategies to improve the reproducibility and performance of the NW integration process. The above strategies aim not only to increase the uniformity of the single-/multi-NW-based electronic layers after the integration process, but also to preserve the as-grown properties of the nanostructures. Moreover, the current approaches to synthesize such a wide selection of organic and inorganic nanostructures, and the availability of specific integration techniques suitable for each kind of nanostructure, reinforce the strategy of using a combination of integration techniques to create heterogeneous structures and systems. For example, metal + semiconductor, organic + inorganic, NTs + NWs, are some combinations of materials that could advance current state-of-the-art LAE applications on both conventional and non-conventional flexible substrates.

References

[1] C. D. Dimitrakopoulos, Organic field-effect transistors for large-area electronics, in Advanced Semiconductor and Organic Nano-Techniques. (Elsevier, 2003), pp. 191–240.

[2] W. T. Navaraj et al., Nanowire FET based neural slement for robotic tactile sensing skin, Frontiers in Neuroscience. 11 (2017), 1–20.

[3] T. Someya et al., A large-area, flexible pressure sensor matrix with organic field-effect transistors for artificial skin applications, Proceedings of the National Academy of Sciences of the United States of America. 101 (2004), 9966–9970.

[4] S. Khan et al., Flexible FETs using ultrathin Si microwires embedded in solution processed dielectric and metal layers, Journal of Micromechanics and Microengineering. 25 (2015), 125019.

[5] S. Hannah et al., Multifunctional sensor based on organic field-effect transistor and ferroelectric poly (vinylidene fluoride trifluoroethylene), Organic Electronics. 56 (2018), 170–177.

[6] S. Khan, L. Lorenzelli and R. Dahiya, Flexible MISFET devices from transfer printed Si microwires and spray coating, IEEE Journal of the Electron Devices Society. 4 (2016), 189–196.

[7] K. Nomura et al., Room-temperature fabrication of transparent flexible thin-film transistors using amorphous oxide semiconductors, Nature. 432 (2004), 488–492.

[8] C. García Núñez et al., Thin film transistors based on zinc nitride as a channel layer for optoelectronic devices, Applied Physics Letters. 101 (2012), 253501.

[9] G. Eda, G. Fanchini and M. Chhowalla, Large-area ultrathin films of reduced graphene oxide as a transparent and flexible electronic material, Nature Nanotechnology. 3 (2008), 270.

[10] M. L. Hammock et al., 25th anniversary article: the evolution of electronic skin (e-skin): a brief history, design considerations, and recent progress, Advanced Materials. 25 (2013), 5997–6038.

[11] C. García Núñez et al., Energy-autonomous, flexible, and transparent tactile skin, Advanced Functional Materials. 27 (2017), 1606287.

[12] T. Sekitani and T. Someya, Stretchable, large-area organic electronics, Advanced Materials. 22 (2010), 2228–2246.

[13] S. Khan et al., Flexible pressure sensors based on screen-printed P (VDF-TrFE) and P (VDF-TrFE)/MWCNTs, IEEE Transactions on Semiconductor Manufacturing. 28 (2015), 486–493.

[14] T. Someya et al., Conformable, flexible, large-area networks of pressure and thermal sensors with organic transistor active matrixes, Proceedings of the National Academy of Sciences of the United States of America. 102 (2005), 12321–12325.

[15] N. Yogeswaran et al., New materials and advances in making electronic skin for interactive robots, Advanced Robotics. 29 (2015), 1359–1373.

[16] C. M. Lampert, Large-area smart glass and integrated photovoltaics, Solar Energy Materials and Solar Cells. 76 (2003), 489–499.

[17] C. García Núñez et al., Large-area self-assembly of silica microspheres/nanospheres by temperature-assisted dip-coating, ACS Applied Materials & Interfaces. 10 (2018), 3058–3068.

[18] M. Simić et al., TiO2-based thick film pH sensor, IEEE Sensors Journal. 17 (2017), 248–255.

[19] W. Dang et al., Stretchable wireless system for sweat pH monitoring, Biosens Bioelectron. 107 (2018), 192–202.

[20] D. Son et al., Multifunctional wearable devices for diagnosis and therapy of movement disorders, Nature Nanotechnology. 9 (2014), 397.

[21] R. S. Dahiya and M. Valle, Robotic Tactile Sensing: Technologies and System. (Springer Science & Business Media, 2012).

[22] I. Kang et al., A carbon nanotube strain sensor for structural health monitoring, Smart materials and structures. 15 (2006), 737.

[23] S. Khan et al., IEEE Transactions on. Flexible Pressure Sensors Based on Screen-Printed P (VDF-TrFE) and P (VDF-TrFE)/MWCNTs, Semiconductor Manufacturing, 28 (2015), 486–493.

[24] L. Manjakkal et al., Printed Flexible Electrochemical pH Sensors based on CuO Nanorods, Sensors and Actuators B: Chemical. 263 (2018), 50–58.

[25] S. Khan, L. Lorenzelli and R. Dahiya, Towards flexible asymmetric MSM structures using Si microwires through contact printing, Semiconductor Science and Technology. 32 (2017), 085013.

[26] F. Bonaccorso et al., Graphene photonics and optoelectronics, Nature Photonics. 4 (2010), 611.

[27] E. O. Polat et al., Synthesis of large area graphene for high performance in flexible optoelectronic devices, Scientific Reports. 5 (2015), 16744.

[28] W. Dang et al., Printable stretchable interconnects, Flexible and Printed Electronics. 2 (2017), 013003.

[29] S. Gupta et al., Ultra-thin chips for high-performance flexible electronics, NPJ Flexible Electronics. DOI: 10.1038/s41528-018-0021-5 (2018).

[30] D.-M. Sun et al., Flexible high-performance carbon nanotube integrated circuits, Nature Nanotechnology. 6 (2011), 156.

[31] A. C. Arias et al., Materials and applications for large area electronics: solution-based approaches, Chem Rev. 110 (2010), 3–24.

[32] C. D. Dimitrakopoulos and P. R. Malenfant, *Or*ganic thin film transistors for large area electronics, Advanced Materials. 14 (2002), 99–117.

[33] A. Javey et al., Layer-by-layer assembly of nanowires for three-dimensional, multifunctional electronics, Nano Letters. 7 (2007), 773–777.

[34] S. G. Rao et al., Nanotube electronics: large-scale assembly of carbon nanotubes, Nature. 425 (2003), 36–37.

[35] S. Khan, L. Lorenzelli and R. S. Dahiya, Technologies for printing sensors and electronics over large flexible substrates: a review, IEEE Sensors Journal. 15 (2015), 3164–3185.

[36] G. Yu, A. Cao and C. M. Lieber, Large-area blown bubble films of aligned nanowires and carbon nanotubes, Nature Nanotechnology. 2 (2007), 372–377.

[37] J. Yao, H. Yan and C. M. Lieber, A nanoscale combing technique for the large-scale assembly of highly aligned nanowires, Nature Nanotechnology. 8 (2013), 329–335.

[38] D. Whang, S. Jin and C. M. Lieber, Large-scale hierarchical organization of nanowires for functional nanosystems, Japanese Journal of Applied Physics. 43 (2004), 4465.

[39] H. Song and M. H. Lee, Combing non-epitaxially grown nanowires for large-area electronic devices, Nanotechnology. 24 (2013), 285302.

[40] A. R. Madaria, A. Kumar and C. Zhou, Large scale highly conductive and patterned transparent films of silver nanowires on arbitrary substrates and their application in touch screens, Nanotechnology. 22 (2011), 245201.

[41] Y.-Z. Long et al., Recent advances in large-scale assembly of semiconducting inorganic nanowires and nanofibers for electronics, sensors and photovoltaics, Chemical Society Reviews. 41 (2012), 4560–4580.

[42] X. Liu et al., Large-scale integration of semiconductor nanowires for high-performance flexible electronics, ACS Nano. 6 (2012), 1888–1900.

[43] Z. Fan et al., Large-scale heterogeneous integration of nanowire arrays for image sensor circuitry, Proceedings of the National Academy of Sciences. 105 (2008), 11066–11070.

[44] X. Duan et al., High-performance thin-film transistors using semiconductor nanowires and nanoribbons, Nature. 425 (2003), 274–278.

[45] M. C. McAlpine et al., Highly ordered nanowire arrays on plastic substrates for ultrasensitive flexible chemical sensors, Nature Materials. 6 (2007), 379–384.

[46] Y. Cui et al., High performance silicon nanowire field effect transistors, Nano Letters. 3 (2003), 149–152.

[47] K. Takei et al., Nanowire active-matrix circuitry for low-voltage macro-scale artificial skin, Nature Materials. 9 (2010), 821–826.

[48] H. Liu et al., Transfer and alignment of random single-walled carbon nanotube films by contact printing, ACS Nano. 4 (2010), 933–938.

[49] D. Roßkopf and S. Strehle, Surface-controlled contact printing for nanowire device fabrication on a large scale, Nanotechnology. 27 (2016), 185301.

[50] Y.-Y. Noh et al., Ink-jet printed ZnO nanowire field effect transistors, Applied Physics Letters. **91** (2007), 043109.

[51] J.-W. Song et al., Inkjet printing of single-walled carbon nanotubes and electrical characterization of the line pattern, Nanotechnology. 19 (2008), 095702.

[52] W. R. Small and M. in het Panhuis, Inkjet printing of transparent electrically conducting single-walled carbon-nanotube composites, Small. 3 (2007), 1500–1503.

[53] R. Yerushalmi et al., Large scale highly ordered assembly of nanowire parallel arrays by differential roll printing, Applied Physics Letters. 91 (2007), 203104.

[54] Z. Fan et al., Wafer-scale assembly of highly ordered semiconductor nanowire arrays by contact printing, Nano letters. 8 (2008), 20–25.

[55] C. García Núñez et al., Conducting properties of nearly depleted ZnO nanowire UV sensors fabricated by dielectrophoresis, Nanotechnology. 24 (2013), 415702.

[56] T. Yu, F.-C. Cheong and C.-H. Sow, The manipulation and assembly of CuO nanorods with line optical tweezers, Nanotechnology. 15 (2004), 1732.

[57] M.A. Bangar et al., Magnetically assembled multisegmented nanowires and their applications, Electroanalysis. 21 (2009), 61–67.

[58] E. K. Hobbie et al., Orientation of carbon nanotubes in a sheared polymer melt, Physics of Fluids. 15 (2003), 1196–1202.

[59] S. Acharya et al., A semiconductor-nanowire assembly of ultrahigh junction density by the Langmuir–Blodgett technique, Advanced Materials. 18 (2006), 210–213.

[60] M. Leitgeb et al., Multilength scale patterning of functional layers by roll-to-roll ultraviolet-light-assisted nanoimprint lithography, ACS Nano. 10 (2016), 4926–4941.

[61] S. H. Ahn and L. J. Guo, *Large-area roll-to-roll and roll-to-plate nanoimprint lithography: a step toward high-throughput application of continuous nanoimprinting*, ACS Nano. 3 (2009), 2304–2310.

[62] H. Schift et al., Nanostructuring of polymers and fabrication of interdigitated electrodes by hot embossing lithography, Microelectronic Engineering. 46 (1999), 121–124.

[63] Q. Xie et al., Fabrication of nanostructures with laser interference lithography, Journal of Alloys and Compounds. 449 (2008), 261–264.

[64] P. Chen et al., Inkjet printing of single-walled carbon nanotube/RuO 2 nanowire supercapacitors on cloth fabrics and flexible substrates, Nano Research. 3 (2010), 594–603.

[65] J. A. Lim et al., Inkjet-printed single-droplet organic transistors based on semiconductor nanowires embedded in insulating polymers, Advanced Functional Materials. 20 (2010), 3292–3297.

[66] S. H. Ko et al., All-inkjet-printed flexible electronics fabrication on a polymer substrate by low-temperature high-resolution selective laser sintering of metal nanoparticles, Nanotechnology. 18 (2007), 345202.

[67] H. Lu et al., Inkjet printed silver nanowire network as top electrode for semi-transparent organic photovoltaic devices, Applied Physics Letters. 106 (2015), 27_21.

[68] M. Law, J. Goldberger and P. Yang, Semiconductor nanowires and nanotubes, Annu. Rev. Mater. Res. 34 (2004), 83–122.

[69] N.P. Dasgupta et al., 25th anniversary article: semiconductor nanowires–synthesis, characterization, and applications, Advanced Materials. 26 (2014), 2137–2184.

[70] C. N. R. Rao et al., Inorganic nanowires, Progress in Solid State Chemistry. 31 (2003), 5–147.

[71] N. Jeon, S. A. Dayeh and L. J. Lauhon, Origin of polytype formation in VLS-grown Ge nanowires through defect generation and nanowire kinking, Nano Letters. 13 (2013), 3947–3952.

[72] J. Hannon et al., The influence of the surface migration of gold on the growth of silicon nanowires, Nature. 440 (2006), 69.

[73] X. Duan and C. M. Lieber, General synthesis of compound semiconductor nanowires, Advanced Materials. 12 (2000), 298–302.

[74] W. Lu and C. M. Lieber, Semiconductor nanowires, Journal of Physics D: Applied Physics. 39 (2006), R387.

[75] B. Bauer et al., Position controlled self-catalyzed growth of GaAs nanowires by molecular beam epitaxy, Nanotechnology. 21 (2010), 435601.

[76] A. Fontcuberta i Morral et al., Nucleation mechanism of gallium-assisted molecular beam epitaxy growth of gallium arsenide nanowires, Applied Physics Letters. 92 (2008), 063112–063112–063113.

[77] C. García Núñez et al., Pure zincblende GaAs nanowires grown by Ga-assisted chemical beam epitaxy, J. Cryst. Growth. 372 (2013), 205–212.

[78] C. Colombo et al., Ga-assisted catalyst-free growth mechanism of GaAs nanowires by molecular beam epitaxy, Physical Review B. 77 (2008), 155326.

[79] C. García Núñez et al., Surface optical phonons in GaAs nanowires grown by Ga-assisted chemical beam epitaxy, Journal of Applied Physics. 115 (2014), 034307.

[80] J. Vukajlovic-Plestin, et al., Engineering the size distributions of ordered GaAs nanowires on silicon, Nano Letters. 17 (2017), 4101–4108.

[81] C. García Núñez et al., GaAs nanowires grown by Ga-assisted chemical beam epitaxy: substrate preparation and growth kinetics, J. Cryst. Growth. 430 (2015), 108–115.

[82] F. Matteini et al., Tailoring the diameter and density of self-catalyzed GaAs nanowires on silicon, Nanotechnology. 26 (2015), 105603.

[83] A. M. Morales and C. M. Lieber, A laser ablation method for the synthesis of crystalline semiconductor nanowires, Science. 279 (1998), 208–211.

[84] C. García Núñez et al., Enhanced fabrication process of zinc oxide nanowires for optoelectronics, Thin Solid Films. 555 (2014), 42–47.

[85] A. I. Persson et al., Solid-phase diffusion mechanism for GaAs nanowire growth, Nature Materials. 3 (2004), 677–681.

[86] K. W. Kolasinski, Catalytic growth of nanowires: vapor–liquid–solid, vapor–solid–solid, solution–liquid–solid and solid–liquid–solid growth, Current Opinion in Solid State and Materials Science. 10 (2006), 182–191.

[87] G. Shen et al., Growth, doping, and characterization of ZnO nanowire arrays, Journal of Vacuum Science & Technology B, Nanotechnology and Microelectronics: Materials, Processing, Measurement, and Phenomena. 31 (2013), 041803.

[88] Z. Huang et al., Metal-assisted chemical etching of silicon: a review, Advanced Materials. 23 (2011), 285–308.

[89] C. Soci et al., ZnO nanowire UV photodetectors with high internal gain, Nano Letters. 7 (2007), 1003–1009.

[90] A. García Marín et al., Fast response ZnO: Al/CuO nanowire/ZnO: Al heterostructure light sensors fabricated by dielectrophoresis, Applied Physics Letters. 102 (2013), 232105.

[91] A. García Marín et al., Continuous-flow system and monitoring tools for the dielectrophoretic integration of nanowires in light sensor arrays, Nanotechnology. 26 (2015), 115502.

[92] C. Thelander et al., Single-electron transistors in heterostructure nano-wires, Applied Physics Letters. 83 (2003), 2052–2054.

[93] L. Tsakalakos et al., Silicon nanowire solar cells, Applied Physics Letters. 91 (2007), 233117.

[94] L. E. Jensen et al., Role of surface diffusion in chemical beam epitaxy of InAs nanowires, Nano Letters. 4 (2004), 1961–1964.

[95] L.-F. Cui et al., Carbon–silicon core–shell nanowires as high capacity electrode for lithium ion batteries, Nano Letters. 9 (2009), 3370–3374.

[96] Y. Wu, R. Fan and P. Yang, Block-by-block growth of single-crystalline Si/SiGe superlattice nanowires, Nano Letters. 2 (2002), 83–86.

[97] P. Caroff et al., *Controlled polytypic and twin-plane superlattices in III–V nanowires*, Nature Nanotechnology. 4 (2009), 50–55.

[98] C. García Núñez et al., *A novel growth method to improve the quality of GaAs nanowires grown by Ga-assisted chemical beam epitaxy*, Nano Letters. 18 (2018), 3608–3615.

[99] C. García Núñez et al., *Heterogeneous integration of contact-printed semiconductor nanowires for high performance devices on large areas*, Microsystems and Nanoengineering. 4 (2018), 22.

[100] C. García Núñez et al., *Effects of hydroxylation and silanization on the surface properties of ZnO nanowires*, ACS Applied Materials & Interfaces. 7 (2015), 5331–5337.

[101] A. I. Persson et al., *Solid-phase diffusion mechanism for GaAs nanowire growth*, Nature Materials. 3 (2004), 677.

[102] J. Wang et al., *Highly polarized photoluminescence and photodetection from single indium phosphide nanowires*, Science. 293 (2001), 1455–1457.

[103] X. Duan et al., *Indium phosphide nanowires as building blocks for nanoscale electronic and optoelectronic devices*, Nature. 409 (2001), 66.

[104] X. Duan and C. M. Lieber, *Laser-assisted catalytic growth of single crystal GaN nanowires*, Journal of the American Chemical Society. 122 (2000), 188–189.

[105] R. Calarco et al., *Nucleation and growth of GaN nanowires on Si (111) performed by molecular beam epitaxy*, Nano Letters. 7 (2007), 2248–2251.

[106] D. Ercolani et al., InAs/InSb nanowire heterostructures grown by chemical beam epitaxy, Nanotechnology. 20 (2009), 505605.

[107] R. LaPierre et al., III–V nanowire photovoltaics: review of design for high efficiency, Physica status solidi (RRL)-Rapid Research Letters. 7 (2013), 815–830.

[108] H. J. Joyce et al., III–V semiconductor nanowires for optoelectronic device applications, Progress in Quantum Electronics. 35 (2011), 23–75.

[109] R. Yan, D. Gargas and P. Yang, Nanowire photonics, Nature Photonics. 3 (2009), 569.

[110] R. Wagner and W. Ellis, Vapor-liquid-solid mechanism of single crystal growth, Applied Physics Letters. 4 (1964), 89–90.

[111] K. A. Dick, A review of nanowire growth promoted by alloys and non-alloying elements with emphasis on Au-assisted III–V nanowires, Progress in Crystal Growth and Characterization of Materials. 54 (2008), 138–173.

[112] K. A. Dick et al., Failure of the vapor– liquid– solid mechanism in Au-assisted MOVPE growth of InAs nanowires, Nano Letters. 5 (2005), 761–764.

[113] H. Xu et al., High-density, defect-free, and taper-restrained epitaxial GaAs nanowires induced from annealed Au thin films, Crystal Growth & Design. 12 (2012), 2018–2022.

[114] S. Breuer et al., Suitability of Au-and self-assisted GaAs nanowires for optoelectronic applications, Nano Letters. 11 (2011), 1276–1279.

[115] E. Russo-Averchi et al., Suppression of three dimensional twinning for a 100% yield of vertical GaAs nanowires on silicon, Nanoscale. 4 (2012), 1486–1490.

[116] S. Hertenberger et al., Growth kinetics in position-controlled and catalyst-free InAs nanowire arrays on Si (111) grown by selective area molecular beam epitaxy, Journal of Applied Physics. 108 (2010), 114316.

[117] M. DeJarld et al., Formation of high aspect ratio GaAs nanostructures with metal-assisted chemical etching, Nano Letters. 11 (2011), 5259–5263.

[118] O. Lourie, D. M. Cox and H. D. Wagner, Buckling and collapse of embedded carbon nanotubes, Physical Review Letters. 81 (1998), 1638–1641.

[119] A. B. Greytak et al., Growth and transport properties of complementary germanium nanowire field-effect transistors, Applied Physics Letters. 84 (2004), 4176–4178.

[120] L. J. Lauhon et al., Epitaxial core–shell and core–multishell nanowire heterostructures, Nature. 420 (2002), 57.

[121] C.-Y. Wen et al., Formation of compositionally abrupt axial heterojunctions in silicon-germanium nanowires, Science. 326 (2009), 1247–1250.

[122] N. Zakharov et al., Growth: growth phenomena of Si and Si/Ge nanowires on Si (1 1 1) by molecular beam epitaxy, J. Cryst. 290 (2006), 6–10.

[123] A. Kołodziejczak-Radzimska and T. Jesionowski, Zinc oxide – from synthesis to application: a review, Materials. 7 (2014), 2833–2881.

[124] Z. L. Wang, Nanostructures of zinc oxide, Materials Today. 7 (2004), 26–33.

[125] J. Cui, Zinc oxide nanowires, Materials Characterization. 64 (2012), 43–52.

[126] D. Shakthivel et al., Growth Mechanisms of 1-D Semiconducting Nanostructures for Flexible and Large Area Electronics. 2018. Cambridge: Cambridge University Press (Cambridge Elements). In press.

[127] M. Hossain et al. Growth of zinc oxide nanowires and nanobelts for gas sensing applications. in Journal of Metastable and Nanocrystalline Materials. 2005. Trans Tech Publ.

[128] J. Zang et al., Tailoring zinc oxide nanowires for high performance amperometric glucose sensor, Electroanalysis. 19 (2007), 1008–1014.

[129] C. Xu et al., Preferential growth of long ZnO nanowire array and its application in dye-sensitized solar cells, The Journal of Physical Chemistry C. 114 (2009), 125–129.

[130] J. C. Johnson et al., Near-field imaging of nonlinear optical mixing in single zinc oxide nanowires, Nano Letters. 2 (2002), 279–283.

[131] Z. L. Wang and J. Song, Piezoelectric nanogenerators based on zinc oxide Nanowire Arrays, Science. 312 (2006), 242–246.

[132] C. Jirayupat et al., Piezoelectric-induced triboelectric hybrid nanogenerators based on the ZnO nanowire layer decorated on the Au/polydimethylsiloxane–Al structure for enhanced triboelectric performance, ACS Applied Materials & Interfaces. 10 (2018), 6433–6440.

[133] Y.-L. Chen et al., The Journal of Physical Chemistry C. *Zinc oxide/ reduced graphene oxide composites and electrochemical capacitance enhanced by homogeneous incorporation of reduced graphene oxide sheets in zinc oxide matrix*, 115 (2011), 2563–2571.

[134] Z. W. Pan, Z. R. Dai and Z. L. Wang, Nanobelts of Semiconducting Oxides, Science. 291 (2001), 1947.

[135] W. I. Park et al., ZnO nanoneedles grown vertically on Si substrates by non-catalytic vapor-phase epitaxy, Adv. Mater. 14 (2002), 1841–1843.

[136] Y. W. Heo et al., Site-specific growth of Zno nanorods using catalysis-driven molecular-beam epitaxy, Applied Physics Letters. 81 (2002), 3046–3048.

[137] J. I. Hong et al., Room-temperature texture-controlled growth of ZnO thin films and their application for growing aligned ZnO nanowire arrays, Nanotechnology. 20 (2009), 5.

[138] R. A. Laudise and A. A. Ballman, Hydrothermal Synthesis of Zinc Oxide and Zinc Sulfide, The Journal of Physical Chemistry. 64 (1960), 688–691.

[139] M. A. Verges, A. Mifsud and C. J. Serna, Formation of rod-like zinc oxide microcrystals in homogeneous solutions, Journal of the Chemical Society, Faraday Transactions. 86 (1990), 959–963.

[140] L. Vayssieres et al., Purpose-built anisotropic metal oxide material: 3D highly oriented microrod array of ZnO, The Journal of Physical Chemistry B. 105 (2001), 3350–3352.

[141] J. Nayak et al., Effect of substrate on the structure and optical properties of ZnO nanorods, Journal of Physics D – Applied Physics. 41 (2008), 6.

[142] E. Givargizov, Fundamental aspects of VLS growth, in Vapour Growth and Epitaxy. 1975, Elsevier. pp. 20–30.

[143] J. Johansson et al., Mass transport model for semiconductor nanowire growth, The Journal of Physical Chemistry B. 109 (2005), 13567–13571.

[144] W. I. Park et al., Metalorganic vapor-phase epitaxial growth of vertically well-aligned ZnO nanorods, Applied Physics Letters. 80 (2002), 4232–4234.

[145] H. T. Ng et al., Optical properties of single-crystalline ZnO nanowires on m-sapphire, Applied Physics Letters. 82 (2003), 2023–2025.

[146] S. Xu and Z. L. Wang, One-dimensional ZnO nanostructures: Solution growth and functional properties, Nano Research. 4 (2011), 1013–1098.

[147] M. N. R. Ashfold et al., The kinetics of the hydrothermal growth of ZnO nanostructures, Thin Solid Films. 515 (2007), 8679–8683.

[148] A. Sugunan et al., Zinc oxide nanowires in chemical bath on seeded substrates: role of hexamine, Journal of Sol-Gel Science and Technology. 39 (2006), 49–56.

[149] S. Xu et al., Density-controlled growth of aligned ZnO nanowire arrays by seedless chemical approach on smooth surfaces, Journal of Materials Research. 23 (2008), 2072–2077.

[150] M. Law, J. Goldberger and P. Yang, Semiconductor Nanowires and Nanotubes, Annual Review of Materials Research. 34 (2004), 83–122.

[151] J. Goldberger et al., Single-crystal gallium nitride nanotubes, Nature. 422 (2003), 599.

[152] R. Fan et al., Fabrication of silica nanotube arrays from vertical silicon nanowire templates, Journal of the American Chemical Society. 125 (2003), 5254–5255.

[153] Q. Wu et al., Synthesis and characterization of faceted hexagonal aluminum nitride nanotubes, Journal of the American Chemical Society. 125 (2003), 10176–10177.

[154] Y. Li, Y. Bando and D. Golberg, Single-crystalline In2O3 nanotubes filled with In, Advanced Materials. 15 (2003), 581–585.

[155] E. Snow et al., High-mobility carbon-nanotube thin-film transistors on a polymeric substrate, Applied Physics Letters. 86 (2005), 033105.

[156] S. Iijima et al., Structural flexibility of carbon nanotubes, The Journal of Chemical Physics. 104 (1996), 2089–2092.

[157] N. Rouhi, D. Jain and P.J. Burke, High-performance semiconducting nanotube inks: Progress and prospects, ACS Nano. 5 (2011), 8471–8487.

[158] Q. Cao and J.A. Rogers, Ultrathin Films of Single-Walled Carbon Nanotubes for Electronics and Sensors: A Review of Fundamental and Applied Aspects, Advanced Materials. 21 (2009), 29–53.

[159] H. J. Dai, Nanotube growth and characterization, Carbon Nanotub Es. 80 (2001), 29–53.

[160] J. Liu et al., Chirality-controlled synthesis of single-wall carbon nanotubes using vapour-phase epitaxy, Nature Communications. 3 (2012), 1199.

[161] B. Liu et al., Chirality-controlled synthesis and applications of single-wall carbon nanotubes, ACS Nano. 11 (2017), 31–53.

[162] Y. Yao et al., "Cloning" of single-walled carbon nanotubes via open-end growth mechanism, Nano Letters. 9 (2009), 1673–1677.

[163] J. R. Sanchez-Valencia et al., Controlled synthesis of single-chirality carbon nanotubes, Nature. 512 (2014), 61.

[164] J. Liu and M. C. Hersam, Recent developments in carbon nanotube sorting and selective growth, MRS Bulletin. 35 (2010), 315–321.

[165] Y. Li et al., Preferential growth of semiconducting single-walled carbon nanotubes by a plasma enhanced CVD method, Nano Letters. 4 (2004), 317–321.

[166] B. Wang et al., Selectivity of single-walled carbon nanotubes by different carbon precursors on Co–Mo catalysts, Journal of the American Chemical Society. 129 (2007), 9014–9019.

[167] D. Ciuparu et al., Uniform-diameter single-walled carbon nanotubes catalytically grown in cobalt-incorporated MCM-41, The Journal of Physical Chemistry B. 108 (2004), 503–507.

[168] N. Izard et al., Semiconductor-enriched single wall carbon nanotube networks applied to field effect transistors, Applied Physics Letters. 92 (2008), 243112.

[169] X. Tu et al., DNA sequence motifs for structure-specific recognition and separation of carbon nanotubes, Nature. 460 (2009), 250.

[170] M. S. Arnold et al., Sorting carbon nanotubes by electronic structure using density differentiation, Nature Nanotechnology. 1 (2006), 60.

[171] A. A. Green and M. C. Hersam, Ultracentrifugation of single-walled nanotubes, Materials Today. 10 (2007), 59–60.

[172] P. G. Collins, M. S. Arnold and P. Avouris, Engineering carbon nano-tubes and nanotube circuits using electrical breakdown, Science. 292 (2001), 706–709.

[173] H. Chen et al., Plasmonic-resonant bowtie antenna for carbon nanotube photodetectors, International Journal of Optics. (2012).

[174] D. S. Hecht et al., Carbon-nanotube film on plastic as transparent electrode for resistive touch screens, Journal of the Society for Information Display. 17 (2009), 941–946.

[175] W. Choi et al., Fully sealed high-brightness carbon-nanotube field-emission display, Applied Physics Letters. 75 (1999), 3129–3131.

[176] Z. Liu et al. Densification of carbon nanotube bundles for interconnect application, in International Interconnect Technology Conference, IEEE 2007. 2007. IEEE.

[177] E. Snow et al., Random networks of carbon nanotubes as an electronic material, Applied Physics Letters. 82 (2003), 2145–2147.

[178] O. Krichevski, E. Tirosh and G. Markovich, Formation of Gold– Silver Nanowires in Thin Surfactant Solution Films, Langmuir. 22 (2006), 867–870.

[179] A. Halder and N. Ravishankar, Ultrafine Single-Crystalline Gold Nanowire Arrays by Oriented Attachment, Advanced Materials. 19 (2007), 1854–1858.

[180] P. Forrer et al., Electrochemical preparation and surface properties of gold nanowire arrays formed by the template technique, Journal of Applied Electrochemistry. 30 (2000), 533–541.

[181] J. H. Song et al., Metal nanowire formation using Mo3Se3-as reducing and sacrificing templates, Journal of the American Chemical Society. 123 (2001), 10397–10398.

[182] C. Wang et al., Ultrathin Au nanowires and their transport properties, Journal of the American Chemical Society. 130 (2008), 8902–8903.

[183] S. Gong et al., A wearable and highly sensitive pressure sensor with ultrathin gold nanowires, Nature Communications. 5 (2014), 3132.

[184] G. Schider et al., Optical properties of Ag and Au nanowire gratings, Journal of Applied Physics. 90 (2001), 3825–3830.

[185] L. Hu et al., Scalable coating and properties of transparent, flexible, silver nanowire electrodes, ACS Nano. 4 (2010), 2955–2963.

[186] A. Tao et al., Langmuir-Blodgett silver nanowire monolayers for mole-cular sensing using surface-enhanced Raman spectroscopy, Nano Letters. 3 (2003), 1229–1233.

[187] T. Kim et al., Uniformly interconnected silver-nanowire networks for transparent film heaters, Advanced Functional Materials. 23 (2013), 1250–1255.

[188] Z. Yu et al., Highly flexible silver nanowire electrodes for shape-memory polymer light-emitting diodes, Advanced Materials. 23 (2011), 664–668.

[189] C. R. Martin, Nanomaterials: a membrane-based synthetic approach, Science. 266 (1994), 1961–1966.

[190] P. M. Ajayan and S. Iijima, Capillarity-induced filling of carbon nanotubes, Nature. 361 (1993), 333–334.

[191] Y. Zhou et al., A novel ultraviolet irradiation photoreduction technique for the preparation of single-crystal Ag nanorods and Ag dendrites, Advanced Materials. e (1999), 850–852.

[192] J.-J. Zhu et al., Preparation of silver nanorods by electrochemical methods, Materials Letters. 49 (2001), 91–95.

[193] D. Zhang et al., Formation of silver nanowires in aqueous solutions of a double-hydrophilic block copolymer, 13 (2001), Chemistry of Materials. 2753–2755.

[194] M. A. Lim et al., A new route toward ultrasensitive, flexible chemical sensors: metal nanotubes by wet-chemical synthesis along sacrificial nanowire templates, ACS Nano. 6 (2011), 598–608.

[195] J. Fu, S. Cherevko and C.-H. Chung, Electroplating of metal nanotubes and nanowires in a high aspect-ratio nanotemplate, Electrochemistry Communications. 10 (2008), 514–518.

[196] Y. Zhang et al., Metal coating on suspended carbon nanotubes and its implication to metal–tube interaction, Chemical Physics Letters. 331 (2000), 35–41.

[197] P. G. Collins and P. Avouris, Nanotubes for electronics, Sci Am. 283 (2000), 62–69.

[198] E. Mutlugun et al., Large-area (over 50 cm × 50 cm) freestanding films of colloidal InP/ZnS quantum dots, Nano Letters. 12 (2012), 3986–3993.

[199] G. I. Koleilat et al., Efficient, stable infrared photovoltaics based on solution-cast colloidal quantum dots, ACS Nano. 2 (2008), 833–840.

[200] X. Dai et al., Solution-processed, high-performance light-emitting diodes based on quantum dots, Nature. 515 (2014), 96.

[201] J.-H. Choi et al., Bandlike transport in strongly coupled and doped quantum dot solids: a route to high-performance thin-film electronics, Nano Letters. 12 (2012), 2631–2638.

[202] C.-S. S. et al., Large-area ordered quantum-dot monolayers via phase separation during spin-casting, Advanced Functional Materials. 15 (2005), 1117–1124.

[203] T.-H. Kim et al., Full-colour quantum dot displays fabricated by transfer printing, Nature Photonics. 5 (2011), 176.

[204] L. Kim et al., Contact printing of quantum dot light-emitting devices, Nano Letters. 8 (2008), 4513–4517.

[205] T.-H. Kim et al., Heterogeneous stacking of nanodot monolayers by dry pick-and-place transfer and its applications in quantum dot light-emitting diodes, Nature Communications. 4 (2013), 2637.

[206] M. K. Choi et al., Wearable red–green–blue quantum dot light-emitting diode array using high-resolution intaglio transfer printing, Nature Communications. 6 (2015), 7149.

[207] V. Wood et al., Inkjet-printed quantum dot–polymer composites for full-color ac-driven displays, Advanced Materials. 21 (2009), 2151–2155.

[208] C. R. Kagan et al., Building devices from colloidal quantum dots, Science. 353 (2016), aac5523.

[209] F. S. Stinner et al., Flexible, high-speed CdSe nanocrystal integrated circuits, Nano Letters. 15 (2015), 7155–7160.

[210] D. K. Kim et al., Flexible and low-voltage integrated circuits constructed from high-performance nanocrystal transistors, Nature Communications. 3 (2012), 1216.

[211] I. J. Kramer and E. H. Sargent, Colloidal quantum dot photovoltaics: a path forward, ACS Nano. 5 (2011), 8506–8514.

[212] G. Konstantatos and E. H. Sargent, Nanostructured materials for photon detection, 5 Nature Nanotechnology. (2010), 391.

[213] K. S. Novoselov et al., Electric field effect in atomically thin carbon films, Science. 306 (2004), 666–669.

[214] Q. H. Wang et al., Electronics and optoelectronics of two-dimensional transition metal dichalcogenides, Nature Nanotechnology. 7 (2012), 699.

[215] K. F. Mak et al., Atomically thin MoS 2: a new direct-gap semiconductor, Physical Review Letters. 105 (2010), 136805.

[216] L. Li et al., Black phosphorus field-effect transistors, Nature Nanotechnology. 9 (2014), 372.

[217] C. R. Dean et al., Boron nitride substrates for high-quality graphene electronics, Nature Nanotechnology. 5 (2010), 722.

[218] S.-J. Han et al., Graphene radio frequency receiver integrated circuit, Nature Communications. 5 (2014), 3086.

[219] K. S. Novoselov et al., A roadmap for graphene, Nature. 490 (2012), 192.

[220] A. B. Sachid et al., Monolithic 3D CMOS using layered semiconductors, Advanced Materials. 28 (2016), 2547–2554.

[221] J. H. An et al., High-performance flexible graphene aptasensor for mercury detection in mussels, ACS Nano. 7 (2013), 10563–10571.

[222] S. Bertolazzi, J. Brivio and A. Kis, Stretching and breaking of ultrathin MoS2, ACS Nano. 5 (2011), 9703–9709.

[223] R. R. Nair et al., Fine structure constant defines visual transparency of graphene, Science. 320 (2008), 1308–1308.

[224] L. Liao et al., High-κ oxide nanoribbons as gate dielectrics for high mobility top-gated graphene transistors, Proceedings of the National Academy of Sciences. 107 (2010), 6711–6715.

[225] X. Li et al., Large-area synthesis of high-quality and uniform graphene films on Copper Foils, Science. 324 (2009), 1312–1314.

[226] J. Cai et al., Atomically precise bottom-up fabrication of graphene nanoribbons, Nature. 466 (2010), 470.

[227] J. N. Coleman et al., Two-dimensional nanosheets produced by liquid exfoliation of layered materials, Science. 331 (2011), 568–571.

[228] Y. Hernandez et al., High-yield production of graphene by liquid-phase exfoliation of graphite, Nature Nanotechnology. 3 (2008), 563.

[229] V. Nicolosi et al., Liquid exfoliation of layered materials, Science. 340 (2013).

[230] Y. Qi et al., Epitaxial graphene on SiC (0001): more than just honeycombs, Physical Review Letters. 105 (2010), 085502.

[231] T. Ohta et al., Controlling the electronic structure of bilayer graphene, Science. 313 (2006), 951–954.

[232] L. Song et al., Large scale growth and characterization of atomic hexagonal boron nitride layers, Nano Letters. 10 (2010), 3209–3215.

[233] S. Sapna et al., Simple fabrication of air-stable black phosphorus heterostructures with large-area hBN sheets grown by chemical vapor deposition method, 2D Materials. 3 (2016), 035010.

[234] C. Chunxiao et al., Synthesis and optical properties of large-area singlecrystalline 2D semiconductor WS2 monolayer from chemical vapor deposition, Advanced Optical Materials. 2 (2014), 131–136.

[235] Z. Yan et al., Toward the Synthesis of Wafer-Scale Single-Crystal Graphene on Copper Foils, ACS Nano. 6 (2012), 9110–9117.

[236] H. Zhou et al., Chemical vapour deposition growth of large single crystals of monolayer and bilayer graphene, Nature Communications. 4 (2013), 2096.

[237] Y. Hao et al., The role of surface oxygen in the growth of large singlecrystal graphene on copper. Science. (2013).

[238] X. Liang et al., Toward clean and crackless transfer of graphene, ACS Nano. 5 (2011), 9144–9153.

[239] M. Her, R. Beams and L. Novotny, Graphene transfer with reduced residue, Physics Letters A. 377 (2013), 1455–1458.

[240] J. Kang et al., Graphene transfer: key for applications, Nanoscale. 4 (2012), 5527–5537.

[241] J. Song et al., A general method for transferring graphene onto soft surfaces, Nat Nano. 8 (2013), 356–362.

[242] S. Bae et al., Roll-to-roll production of 30-inch graphene films for transparent electrodes, Nature Nanotechnology. 5 (2010), 574–578.

[243] J. Kang et al., Efficient transfer of large-area graphene films onto rigid substrates by hot pressing, ACS Nano. 6 (2012), 5360–5365.

[244] J. Chen et al., Oxygen-aided synthesis of polycrystalline graphene on silicon dioxide substrates, Journal of the American Chemical Society. 133 (2011), 17548–17551.

[245] F. Qingliang et al., Growth of large-area 2D MoS2(1-x)Se2x semiconductor alloys, Advanced Materials. 26 (2014), 2648–2653.

[246] J.-H. Lee et al., Wafer-scale growth of single-crystal monolayer graphene on reusable hydrogen-terminated germanium, Science. 344 (2014), 286–289.

[247] L. Gao et al., *Face-to-face transfer of wafer-scale graphene films*, Nature. 505 (2013), 190.

[248] S. Rahimi et al., Toward 300 mm wafer-scalable high-performance polycrystalline chemical vapor deposited graphene transistors, ACS Nano. 8 (2014), 10471–10479.

[249] L. Yu et al., Design, modeling, and fabrication of chemical vapor deposition grown MoS2 circuits with E-mode FETs for large-area electronics, Nano Letters. 16 (2016), 6349–6356.

[250] K. S. Kim et al., Large-scale pattern growth of graphene films for stretchable transparent electrodes, Nature. 457 (2009), 706.

[251] C. Hsiao-Yu et al., Large-area monolayer MoS2 for flexible low-power RF nanoelectronics in the GHz regime, Advanced Materials. 28 (2016), 1818–1823.

[252] N. Liu et al., Large-area, transparent, and flexible infrared photodetector fabricated using PN junctions formed by N-doping chemical vapor deposition grown graphene, Nano Letters. 14 (2014), 3702–3708.

[253] D. De Fazio et al., High responsivity, large-area graphene/MoS2 flexible photodetectors, ACS Nano. 10 (2016), 8252–8262.

[254] C. G. Núñez et al., Energy-autonomous, flexible, and transparent tactile skin, Advanced Functional Materials. 27 (2017), 1606287-n/a.

[255] H. Lee et al., A graphene-based electrochemical device with thermo-responsive microneedles for diabetes monitoring and therapy, Nature Nanotechnology. 11 (2016), 566.

[256] S. Wagner, et al., Highly sensitive electromechanical piezoresistive pressure sensors based on large-area layered PtSe2 films, Nano Letters. **18** (2018), 3738–3745.

[257] M. F. El-Kady et al., Laser scribing of high-performance and flexible graphene-based electrochemical capacitors, Science. 335 (2012), 1326–1330.

[258] A. Scidà et al., Application of graphene-based flexible antennas in consumer electronic devices, Materials Today. 21 (2018), 223–230.

[259] D. Voiry et al., High-quality graphene via microwave reduction of solution-exfoliated graphene oxide, Science. 353 (2016), 1413–1416.

[260] J. H. Oh et al., Solution-processed, high-performance n-channel organic microwire transistors, Proceedings of the National Academy of Sciences. 106 (2009), 6065–6070.

[261] T. J. Shin et al., Tunable thin-film crystalline structures and field-effect mobility of oligofluorene–thiophene derivatives, Chemistry of Materials. 19 (2007), 5882–5889.

[262] A. L. Briseno et al., Introducing organic nanowire transistors, Materials Today. 11 (2008), 38–47.

[263] C. Kloc et al., Physical vapor growth of centimeter-sized crystals of α-hexathiophene, J. Cryst. Growth. 182 (1997), 416–427.

[264] D. Li and Y. Xia, Electrospinning of nanofibers: reinventing the wheel?, Advanced Materials. 16 (2004), 1151–1170.

[265] S.-Y. Min et al., Large-scale organic nanowire lithography and electronics, Nature Communications. 4 (2013), 1773.

[266] Q. Tang, et al., Organic nanowire crystals combine Excellent Device Performance and Mechanical Flexibility, Small. **7** (2011), 189–193.

[267] T. Sekitani et al., Flexible organic transistors and circuits with extreme bending stability, Nature Materials. 9 (2010), 1015.

[268] Y. S. Zhao, J. Wu and J. Huang, Vertical organic nanowire arraysc: controlled synthesis and chemical sensors, Journal of the American Chemical Society. 131 (2009), 3158–3159.

[269] X. Wang et al., High-performance organic–inorganic hybrid photodetectors based on P3HT:CdSe nanowire heterojunctions on rigid and flexible substrates, Advanced Functional Materials. 23 (2013), 1202–1209.

[270] T. Sekitani et al., Organic nonvolatile memory transistors for flexible sensor arrays, Science. 326 (2009), 1516–1519.

[271] J. D. Park, S. Lim and H. Kim, Patterned silver nanowires using the gravure printing process for flexible applications, Thin Solid Films. 586 (2015), 70–75.

[272] Q. Qi et al., Properties of humidity sensing ZnO nanorods-base sensor fabricated by screen-printing, Sensors and Actuators B: Chemical. 133 (2008), 638–643.

[273] L. V. Thong et al., On-chip fabrication of SnO2-nanowire gas sensor: the effect of growth time on sensor performance, Sensors and Actuators B: Chemical. 146 (2010), 361–367.

[274] J. Wang and M. Musameh, Carbon nanotube screen-printed electrochemical sensors, Analyst. 129 (2004), 1–2.

[275] J. Liang, K. Tong and Q. Pei, A water-based silver-nanowire screen-print ink for the fabrication of stretchable conductors and wearable thin-film transistors, Advanced Materials. 28 (2016), 5986–5996.

[276] D. Li, Y. Wang and Y. Xia, Electrospinning of polymeric and ceramic nanofibers as uniaxially aligned arrays, Nano Letters. 3 (2003), 1167–1171.

[277] Q. Zhang et al., Electrospun carbon nanotube composite nanofibres with uniaxially aligned arrays, Nanotechnology. 18 (2007), 115611.

[278] D. Li, T. Herricks and Y. Xia, Magnetic nanofibers of nickel ferrite prepared by electrospinning, Applied Physics Letters. 83 (2003), 4586–4588.

[279] C. Lai et al., Effects of humidity on the ultraviolet nanosensors of aligned electrospun ZnO nanofibers, RSC Advances. 3 (2013), 6640–6645.

[280] H.-W. Lu et al., Fabrication and electrochemical properties of three-dimensional net architectures of anatase TiO2 and spinel Li4Ti5O12 nanofibers, Journal of Power Sources. 164 (2007), 874–879.

[281] D. Li and Y. Xia, Direct fabrication of composite and ceramic hollow nanofibers by electrospinning, Nano Letters. 4 (2004), 933–938.

[282] S. H. Choi et al., Hollow ZnO nanofibers fabricated using electrospun polymer templates and their electronic transport properties, ACS Nano. 3 (2009), 2623–2631.

[283] C. L. Pint et al., Formation of highly dense aligned ribbons and transparent films of single-walled carbon nanotubes directly from carpets, ACS Nano. 2 (2008), 1871–1878.

[284] T. Takahashi et al., Monolayer resist for patterned contact printing of aligned nanowire arrays, Journal of the American Chemical Society. 131 (2009), 2102–2103.

[285] K. W. Min et al., White-light-emitting diode array of p+-Si/Aligned n-SnO2 nanowires heterojunctions, Advanced Functional Materials. 21 (2011), 119–124.

[286] J. Im et al., Direct printing of aligned carbon nanotube patterns for high-performance thin film devices, Applied Physics Letters. 94 (2009), 053109.

[287] S. Li et al., Transfer printing of submicrometer patterns of aligned carbon nanotubes onto functionalized electrodes, Small. 3 (2007), 616–621.

[288] M. A. Meitl et al., Transfer printing by kinetic control of adhesion to an elastomeric stamp, Nat Mater. 5 (2006), 33–38.

[289] A. Javey et al., Ballistic carbon nanotube field-effect transistors, Nature. 424 (2003), 654–657.

[290] A. Javey et al., High-κ dielectrics for advanced carbon-nanotube transistors and logic gates, Nature Materials. 1 (2002), 241–246.

[291] E. Joselevich and C. M. Lieber, Vectorial growth of metallic and semiconducting single-wall carbon nanotubes, Nano Letters. 2 (2002), 1137–1141.

[292] C. Kocabas et al., Aligned arrays of single-walled carbon nanotubes generated from random networks by orientationally selective laser ablation, Nano Letters. 4 (2004), 2421–2426.

[293] S. Huang et al., Growth mechanism of oriented long single walled carbon nanotubes using "fast-heating" chemical vapor deposition process, Nano Letters. 4 (2004), 1025–1028.

[294] S. J. Kang et al., High-performance electronics using dense, perfectly aligned arrays of single-walled carbon nanotubes, Nature Nanotechnology. 2 (2007), 230–236.

[295] Q. Cao et al., Highly bendable, transparent thin-film transistors that use carbon-nanotube-based conductors and semiconductors with elastomeric dielectrics, Advanced Materials. 18 (2006), 304–309.

[296] Y. K. Chang and F. C. Hong, The fabrication of ZnO nanowire field-effect transistors by roll-transfer printing, Nanotechnology. 20 (2009), 195302.

[297] F. Xu and Y. Zhu, Highly conductive and stretchable silver nanowire conductors, Advanced Materials. 24 (2012), 5117–5122.

[298] C.-M. Hsu et al., Wafer-scale silicon nanopillars and nanocones by Langmuir–Blodgett assembly and etching, Applied Physics Letters. 93 (2008), 133109.

[299] M. Szekeres et al., Ordering and optical properties of monolayers and multilayers of silica spheres deposited by the Langmuir–Blodgett method, Journal of Materials Chemistry. 12 (2002), 3268–3274.

[300] A. Ulman, An Introduction to Ultrathin Organic Films: From Langmuir–Blodgett to Self–Assembly. 2013. Academic Press.

[301] D. Wang et al., Oxidation resistant germanium nanowires: bulk synthesis, long chain alkanethiol functionalization, and Langmuir-Blodgett assembly, Journal of the American Chemical Society. **127** (2005), 11871–11875.

[302] C. A. Stover, D. L. Koch and C. Cohen, Observations of fibre orientation in simple shear flow of semi-dilute suspensions, Journal of Fluid Mechanics. **238** (1992), 277–296.

[303] D. Maria and B. Peter, Dielectrophoresis of carbon nanotubes using microelectrodes: a numerical study, Nanotechnology. **15** (2004), 1095.

[304] S. Raychaudhuri et al., Precise semiconductor nanowire placement through dielectrophoresis, Nano Letters. **9** (2009), 2260–2266.

[305] J. Suehiro, G. Zhou and M. Hara, Fabrication of a carbon nanotube-based gas sensor using dielectrophoresis and its application for ammonia detection by impedance spectroscopy, Journal of Physics D: Applied Physics. **36** (2003), L109.

[306] D. Wang et al., Controlled assembly of zinc oxide nanowires using dielectrophoresis, Applied Physics Letters. **90** (2007), 103110.

[307] C. García Núñez et al. Photodetector fabrication by dielectrophoretic assembly of GaAs nanowires grown by a two-steps method, in SPIE Nanoscience + Engineering. 2017. San Diego: SPIE.

[308] M. Liu et al., Self-assembled magnetic nanowire arrays, Applied Physics Letters. **90** (2007), 103105.

[309] A. K. Salem, et al., Receptor-mediated self-assembly of multi-component magnetic nanowires, Advanced Materials. 16 (2004), 268–271.

[310] M. Tanase et al., Assembly of multicellular constructs and microarrays of cells using magnetic nanowires, Lab on a Chip. 5 (2005), 598–605.

[311] R. C. Gauthier, M. Ashman and C. P. Grover, Experimental confirmation of the optical-trapping properties of cylindrical objects, Applied Optics. 38 (1999), 4861–4869.

[312] A. Irrera et al., Size-scaling in optical trapping of silicon nanowires, Nano Letters. 11 (2011), 4879–4884.

[313] P.J. Pauzauskie et al., Optical trapping and integration of semiconductor nanowire assemblies in water, Nature Materials. 5 (2006), 97–101.

[314] Z. Yan et al., Three-dimensional optical trapping and manipulation of single silver nanowires, Nano Letters. 12 (2012), 5155–5161.

[315] K. Cantor, Blown Film Extrusion. 2011. Hanser Publications.

[316] Y. Huang et al., Directed assembly of one-dimensional nanostructures into functional networks, Science. 291 (2001), 630–633.

[317] D. Whang et al., Large-scale hierarchical organization of nanowire arrays for integrated nanosystems, Nano Letters. 3 (2003), 1255–1259.

[318] D. Whang, S. Jin and C. M. Lieber, Nanolithography using hierarchically assembled nanowire masks, Nano Letters. 3 (2003), 951–954.

[319] O. Albrecht et al., Construction and use of LB deposition machines for pilot production, Thin Solid Films. 284 (1996), 152–156.

[320] M. Burghard et al., Controlled adsorption of carbon nanotubes on chemically modified electrode arrays, Advanced Materials. 10 (1998), 584–+.

[321] P. A. Smith et al., Electric-field assisted assembly and alignment of metallic nanowires, Applied Physics Letters. 77 (2000), 1399–1401.

[322] M. C. Wang et al., Electrokinetic assembly of selenium and silver nanowires into macroscopic fibers, ACS Nano. 4 (2010), 2607–2614.

[323] C. H. Lee, D. R. Kim and X. Zheng, Orientation-controlled alignment of axially modulated pn silicon nanowires, Nano Letters. 10 (2010), 5116–5122.

[324] X. Duan et al., Indium phosphide nanowires as building blocks for nanoscale electronic and optoelectronic devices, Nature. 409 (2001), 66–69.

[325] R. Zhou et al., CdSe nanowires with illumination-enhanced conductivity: Induced dipoles, dielectrophoretic assembly, and field-sensitive emission, Journal of Applied Physics. 101 (2007), 073704.

[326] A. O'Riordan et al., Dielectrophoretic self-assembly of polarized light emitting poly (9, 9-dioctylfluorene) nanofibre arrays, Nanotechnology. 22 (2011), 105602.

[327] Y. Dan et al., Dielectrophoretically assembled polymer nanowires for gas sensing, Sensors and Actuators B: Chemical. 125 (2007), 55–59.

[328] A. Motayed et al., Realization of reliable GaN nanowire transistors utilizing dielectrophoretic alignment technique, Journal of Applied Physics. 100 (2006), 114310.

[329] A. D. Wissner-Gross, Dielectrophoretic reconfiguration of nanowire interconnects, Nanotechnology. 17 (2006), 4986.

[330] S. Evoy et al., Dielectrophoretic assembly and integration of nanowire devices with functional CMOS operating circuitry, Microelectronic Engineering. 75 (2004), 31–42.

[331] A. Jamshidi et al., Dynamic manipulation and separation of individual semiconducting and metallic nanowires, Nature Photonics. 2 (2008), 86–89.

[332] Y. Liu et al., Dielectrophoretic assembly of nanowires, The Journal of Physical Chemistry B. 110 (2006), 14098–14106.

[333] J. Suehiro et al., Dielectrophoretic fabrication and characterization of a ZnO nanowire-based UV photosensor, Nanotechnology. 17 (2006), 2567.

[334] T. B. Jones and T. B. Jones, Electromechanics of Particles. 2005. Cambridge University Press.

[335] S. Myung et al., Large-scale "surface-programmed assembly" of pristine vanadium oxide nanowire-based devices, Advanced Materials. 17 (2005), 2361–2364.

[336] J. Kang et al., Massive assembly of ZnO nanowire-based integrated devices, Nanotechnology. 19 (2008), 095303.

[337] M. Chen and P. C. Searson, The dynamics of nanowire self-assembly, Advanced Materials. 17 (2005), 2765–2768.

[338] J. Lee et al., DNA assisted assembly of multisegmented nanowires, Electroanalysis. 19 (2007), 2287–2293.

[339] A. Bachtold et al., Logic circuits with carbon nanotube transistors, Science. 294 (2001), 1317–1320.

[340] U. Bockelmann et al., Unzipping DNA with optical tweezers: high sequence sensitivity and force flips, Biophysical Journal. 82 (2002), 1537–1553.

[341] R. Agarwal et al., Manipulation and assembly of nanowires with holographic optical traps, Opt Express. 13 (2005), 8906–8912.

[342] Q. Tang et al., Assembly of Nanoscale Organic Single-Crystal Cross-Wire Circuits, Advanced Materials. 21 (2009), 4234–4237.

[343] A. Ashkin, J. M. Dziedzic and T. Yamane, Optical trapping and manipulation of single cells using infrared laser beams, Nature. 330 (1987), 769.

[344] Y. Pang and R. Gordon, Optical trapping of a single protein, Nano Letters. 12 (2011), 402–406.

[345] L. Nugent-Glandorf and T. T. Perkins, Measuring 0.1-nm motion in 1 ms in an optical microscope with differential back-focal-plane detection, Optics Letters. 29 (2004), 2611–2613.

[346] W. A. Shelton, K. D. Bonin and T. G. Walker, Nonlinear motion of optically torqued nanorods, Physical Review E. 71 (2005), 036204.

[347] K. Takei et al., Nanowire active-matrix circuitry for low-voltage macro-scale artificial skin, Nature Materials. 9 (2010), 821.

[348] T. Takahashi et al., Contact printing of compositionally graded CdSxSe1− x nanowire parallel arrays for tunable photodetectors, Nanotechnology. 23 (2012), 045201.

[349] M. J. Allen et al., Soft transfer printing of chemically converted graphene, Advanced Materials. 21 (2009), 2098–2102.

[350] Q. Chen, C. Martin and D. Cumming, Transfer printing of nanoplasmonic devices onto flexible polymer substrates from a rigid stamp, Plasmonics. 7 (2012), 755–761.

[351] Y.-K. Chang and F. C.-N. Hong, The fabrication of ZnO nanowire field-effect transistors by roll-transfer printing, Nanotechnology. 20 (2009), 195302.

[352] J. Yao, H. Yan and C. M. Lieber, A nanoscale combing technique for the large-scale assembly of highly aligned nanowires, Nature Nanotechnology. 8 (2013), 329.

[353] S. Wu et al., Blown bubble assembly of graphene oxide patches for transparent electrodes in carbon–silicon solar cells, ACS Applied Materials & Interfaces. 7 (2015), 28330–28336.

[354] S. Wu et al., Soluble polymer-based, blown bubble assembly of single- and double-layer nanowires with shape control, ACS Nano. 8 (2014), 3522–3530.

[355] A. Tao et al., Langmuir–Blodgett silver nanowire monolayers for molecular sensing using surface-enhanced Raman spectroscopy, Nano Letters. 3 (2003), 1229–1233.

[356] A. Gang et al., Microfluidic alignment and trapping of 1D nanostructures – a simple fabrication route for single-nanowire field effect transistors, RSC Advances. 5 (2015), 94702–94706.

[357] X. Li, O. Niitsoo and A. Couzis, *Electrostatically assisted fabrication of silver–dielectric core/shell nanoparticles thin film capacitor with uniform metal nanoparticle distribution and controlled spacing*, Journal of Colloid and Interface Science. 465 (2016), 333–341.

[358] P. Parthangal, R. Cavicchi and M. Zachariah, Design, fabrication and testing of a novel gas sensor utilizing vertically aligned zinc oxide nanowire arrays, MRS Online Proceedings Library Archive. 951 (2006).

[359] C. L. Zhang et al., Macroscopic-scale alignment of ultralong Ag nanowires in polymer nanofiber mat and their hierarchical structures by magnetic-field-assisted electrospinning, Small. 8 (2012), 2936–2940.

[360] Y. Huang et al., Magnetic-assisted, self-healable, yarn-based supercapacitor, ACS Nano. 9 (2015), 6242–6251.

[361] M.-S. Kim et al., Flexible conjugated polymer photovoltaic cells with controlled heterojunctions fabricated using nanoimprint lithography, Applied Physics Letters. 90 (2007), 123113.

[362] J. J. Wang et al., 30-nm-wide aluminum nanowire grid for ultrahigh contrast and transmittance polarizers made by UV-nanoimprint lithography, Applied Physics Letters. 89 (2006), 141105.

[363] C.-J. Ting et al., Low cost fabrication of the large-area anti-reflection films from polymer by nanoimprint/hot-embossing technology, Nanotechnology. 19 (2008), 205301.

[364] N. S. Cameron et al., High fidelity, high yield production of microfluidic devices by hot embossing lithography: rheology and stiction, Lab on a Chip. 6 (2006), 936–941.

[365] J. De Boor et al., Sub-100 nm silicon nanowires by laser interference lithography and metal-assisted etching, Nanotechnology. 21 (2010), 095302.

[366] L. J. Guo, Nanoimprint lithography: methods and material requirements, Advanced Materials. 19 (2007), 495–513.

[367] W. Zhang and S. Y. Chou, Fabrication of 60-nm transistors on 4-in. wafer using nanoimprint at all lithography levels, Applied Physics Letters. 83 (2003), 1632–1634.

[368] D. Pisignano et al., Room-temperature nanoimprint lithography of non-thermoplastic organic films, Advanced Materials. 16 (2004), 525–529.

[369] L. J. Guo, X. Cheng and C.-F. Chou, Fabrication of size-controllable nanofluidic channels by nanoimprinting and its application for DNA stretching, Nano Letters. 4 (2004), 69–73.

[370] C. Pina-Hernandez et al., High-throughput and etch-selective nanoimprinting and stamping based on fast-thermal-curing poly(dimethylsiloxane)s, Advanced Materials. 19 (2007), 1222–1227.

[371] S. H. Ahn, J.-S. Kim and L. J. Guo, Bilayer metal wire-grid polarizer fabricated by roll-to-roll nanoimprint lithography on flexible plastic substrate, Journal of Vacuum Science & Technology B: Microelectronics and Nanometer Structures Processing, Measurement, and Phenomena. 25 (2007), 2388–2391.

[372] H. Tan, A. Gilbertson and S. Y. Chou, Roller nanoimprint lithography, Journal of Vacuum Science & Technology B: Microelectronics and Nanometer Structures Processing, Measurement, and Phenomena. 16 (1998), 3926–3928.

[373] S.-M. Seo, T.-I. Kim and H. H. Lee, Simple fabrication of nanostructure by continuous rigiflex imprinting, Microelectronic Engineering. 84 (2007), 567–572.

[374] S. H. Ahn and L. J. Guo, High-speed roll-to-roll nanoimprint lithography on flexible plastic substrates, Advanced Materials. 20 (2008), 2044–2049.

[375] C. David and D. Hambach, Line width control using a defocused low voltage electron beam, Microelectronic Engineering. 46 (1999), 219–222.

[376] P. R. Krauss and S. Y. Chou, Nano-compact disks with 400 Gbit/in 2 storage density fabricated using nanoimprint lithography and read with proximal probe, Applied Physics Letters. 71 (1997), 3174–3176.

[377] J. H. Jang et al., 3D micro-and nanostructures via interference lithography, Advanced Functional Materials. 17 (2007), 3027–3041.

[378] W. Choi et al., Synthesis of silicon nanowires and nanofin arrays using interference lithography and catalytic etching, Nano Letters. 8 (2008), 3799–3802.

Acknowledgements

This work was supported by EPSRC Engineering Fellowship for Growth – Printable Tactile Skin (EP/M002527/1), EPSRC Programme Grant – Heteroprint (EP/R03480X/1) and Lord Kelvin Adam Smith Fellowship by University of Glasgow.

Cambridge Elements ≡

Flexible and Large-Area Electronics

Ravinder Dahiya
University of Glasgow

Ravinder Dahiya is Professor of Electronic and Nanoscale Engineering, and an EPSRC Fellow, at the University of Glasgow, where he leads the Bendable Electronics and Sensing Technologies (BEST) group. He is a Distinguished Lecturer of the IEEE Sensors Council and serves on the Editorial Boards of *Scientific Reports*, the *IEEE Sensors Journal* and *IEEE Transactions on Robotics*. He is an expert in the field of flexible and bendable electronics and electronic skin.

Luigi G. Occhipinti
University of Cambridge

Luigi G. Occhipinti is a Principal Investigator in the Department of Engineering at the University of Cambridge and Deputy Director and COO of the Cambridge Graphene Centre. He is CEO at Cambridge Innovation Technologies Consulting Ltd and Non-Executive Director of Zinergy UK Ltd. He is a recognised expert in printed, organic and large-area electronics and integrated smart systems with over 20 years' experience in the semiconductor industry, and is a former R&D Senior Group Manager and Programs Director at STMicroelectronics.

About the Series

This innovative series provides authoritative coverage of the state-of-the-art in bendable and large-area electronics. Specific Elements provide in-depth coverage of key technologies, materials and techniques for the design and manufacturing of flexible electronic circuits and systems, as well as cutting-edge insights into emerging real-world applications. This series is a dynamic reference resource for graduate students, researchers and practitioners in electrical engineering, physics, chemistry and materials.

Cambridge Elements $\equiv$

Flexible and Large-Area Electronics

Elements in the Series

Bioresorbable Materials and Their Application in Electronics
Xian Huang
9781108406239

Organic and Amorphous-Metal-Oxide Flexible Analogue Electronics
Pecunia et al.
9781108458191

Integration Techniques for Micro/Nanostructure-Based Large-Area Electronics
García Núñez et al.
9781108703529

A full series listing is available at: www.cambridge.org/eflex

Printed in the United States
By Bookmasters